Fossilization

SCIENCES

Geoscience, Field Director – Yves Lagabrielle

The Surface of the Lithosphere and the Biosphere,
Subject Head – François Baudin

Fossilization

Fundamental Principles and Cases of Exceptional Fossilizations

Didier Néraudeau
Jean-David Moreau

iSTE

WILEY

First published 2025 in Great Britain and the United States by ISTE Ltd and John Wiley & Sons, Inc.

ISTE Ltd
27-37 St George's Road
London SW19 4EU
UK

www.iste.co.uk

John Wiley & Sons, Inc.
111 River Street
Hoboken, NJ 07030
USA

www.wiley.com

Library of Congress Control Number: 2025939529

British Library Cataloguing-in-Publication Data
A CIP record for this book is available from the British Library
ISBN 978-1-78945-212-9

ERC code:
PE10 Earth System Science
PE10_12 Sedimentology, soil science, palaeontology, earth evolution

The manufacturer's authorized representative according to the EU General Product Safety Regulation is Wiley-VCH GmbH, Boschstr. 12, 69469 Weinheim, Germany, e-mail: Product_Safety@wiley.com.

Contents

Foreword

Bruno DAVID
Anciennement Muséum National d'Histoire Naturelle, Paris, France

"You old fossil!" This expression is, in itself, a little outdated, but aims to highlight that its target is over the hill, old fashioned, a relic from a bygone age. Is that all that can be said on the matter? Perhaps not.

The first question that springs to mind is this: is a fossil necessarily old? Well, on a human time scale, certainly it is, but in terms of geological time, which is thought of in millions of years, *it ain't necessarily so*. In order to answer such a question, we need to think about the very process of fossilization, and the time that it takes to occur. When I say "*the* process", I am oversimplifying, because in reality, it is process*es*, plural – multiple and very different pathways – by which a set of remains become fossilized.

The second question relates to the term "fossil": is it unequivocal, relating only to a certain type of object? Not at all: fossils are very diverse, on two fronts. The first is the nature of what is being fossilized. Commonly, we might find an ammonite from the Jurassic period, a shark's tooth from the Tertiary period, a fern leaf from the Carboniferous period, etc. Rarer, or at least less well known, we might find a dinosaur footprint or egg, a coprolite – that is, fossilized dung – spores or microscopic pollen, etc. In exceptional cases, we might see preserved embryos, feathers which still have traces of their original coloring, decorative patterns on mollusk shells, insects almost completely intact, and so on. Next, fossils can be said to be diverse in terms of the means of fossilization: silicification, which preserves the finest of details; pyritization, which makes the fossil sparkle; inclusion in transparent amber; phosphatization; or, more typically, calcification. Mix these two factors of diversity in fossils, stir them well, and you will find a breathtaking range of sizes, shapes, patterns, colors, etc. In this marvelous collection, though, chaos appears to reign.

Didier Néraudeau and Jean-David Moreau's book takes the reader by the hand, and masterfully demonstrates that this apparent chaos is merely illusive, that there is method in the apparent madness, that sense and order can be found within it. For this purpose, they begin with the processes of fossilization, then detail the types of mineralization which, as you will see, are more numerous than you could possibly imagine. Finally, they offer an irresistible insight into the contexts that have led to a number of the most remarkable instances of fossilization, with even individual molecules having been fossilized. Everything falls into place, with the flourishing diversity fitting into boxes that are primarily formed by the physical-chemical context. Readers will be left with the unfaltering understanding that all fossils are exceptional – even those which seem most ordinary.

The interjection with which this foreword opens is intended to be disparaging, with a fossil in this instance being perceived as something holding little interest – a lowly thing. Nothing could be further from the truth! Thus, next time you hear the words "You old fossil!" – particularly if it is aimed at you – take it as a compliment, because even the most lowly fossil is an exceptional thing, the result of a rare and complex process. Often, they are something quite beautiful, as demonstrated by the superb illustrations supplied in this book; as a result, fossils have always been objects of fascination, sometimes cult items, thought to be imbued with amazing properties. I have no doubt that as you read this book, the way in which you see fossils will be refined – transformed, even – and that is the highest praise I can bestow upon this book.

Acknowledgments

The authors would like to thank the following people and organizations for their contributions to the illustrations of this book: C. Bara (Musée Laval), R.-A. Benoit (Musée Laval), S. Beurel, T. Bouyer, M. Brazidec (Université de Rennes), L. Cazes, S. Charbonnier (MNHN, Paris), P. Courville (Université de Rennes), B. Coutret, T. Danelian (Université de Lille), P. DeWever (MNHN, Paris), D. Desmarres (Sorbonne Université), É. Dépré, S. Gardin (CNRS, Sorbonne Université), D. Gendry (Université de Rennes), R. Gougeon (Université de Bretagne Occidentale), D. Grosheny (Université de Strasbourg), M. Guillo (MG photographie, Locminé), B. Hogrel, R. Jattiot (CNRS, Sorbonne Université), M. Jolivet (Université de Rennes), C. Jouault (MNHN, Paris), P. Lacroix, M. Le Couls (Université de Rennes), E. Lasseur (BRGM, Orléans), X. Le Coz (Université de Rennes), T. Le Diouron, P. Loubry, P. Manicart, E. Masure (Sorbonne Université), B. Mellier (Musée Angers), P. Miramand (La Rochelle Université), N. Morel (Musée Vert, Le Mans), M. Ollivier (Université de Rennes), F. Paris (CNRS, Université de Rennes), V. Perrichot (Université de Rennes), M. Philippe (Université Claude Bernard, Lyon), N. Podevigne (Sorbonne Université), F. Polette (Université de Rennes), A. Radix

(Musée de Bourges), N. Robin (CNRS, Université de Rennes), J.-P. Saint Martin (MNHN, Paris), S. Saint Martin (MNHN, Paris), T. Servais (Université de Lille), Ch. Strullu (Natural History Museum, London), P. Tassy (MNHN, Paris), A. Trébaol (Université de Rennes), X. Valentin (Université de Poitiers), R. Vullo (CNRS, Université de Rennes), and R. Zurbach (Réserve géologique du Lubéron).

We thank the private collectors and public institutions for allowing us to use images from their collections: Amy Beauvois, Kevin Chartier, Éric Dépré, Damien Gendry and the collections of the Institut de géologie de Rennes (IGR), Bernard Hogrel, Gérard Lafaurie, Jean-Claude Le Sager, Jenny Bramail and the Australian Opal Centre, the Association paléontologique des hauts plateaux du Languedoc (APHPL), Musée de Bourges, the Musée de Minéralogie et de Paléontologie du Bugue, the Musée de Lodève, the Muséum d'Histoire naturelle de Paris, the Muséum de La Rochelle, the Musée d'Angoulême, the Maison de la minéralogie du Massif central (Antignac), the Beijing Institute of Geology and Paleontology, the Réserve géologique du Lubéron, the universities of Bourgogne, Lille, Lyon I, Rennes 1, Aberdeen and Gongju (South Korea).

We are also grateful to the European Synchrotron Radiation Facility (ESRF, Grenoble) for supplying the tomography images presented in this book.

Last but not least, warm thanks go to Bruno David (MNHN, Paris) for agreeing to write the foreword for this book, with a touch of humor and the expert eye of a renowned paleobiologist.

1

General Points

1.1. Fossilization: a phenomenon that is always exceptional

In most ecosystems, dead organisms decompose and disappear in relatively little time. To begin with, the action of decomposers and necrophages breaks down the soft tissues of the bodies, and these organic remains are therefore not protected by being buried in the sediments. Secondly, the remains which are most mineralized – bones, teeth and shells, for example – are subject to bioerosion (by encrusting organisms, perforators, crushing carrion eaters, etc.) and physical fragmentation (due to transportation in water, temperature variations, etc.), and are gradually reduced to debris, eroded to a greater or lesser extent. The time needed for fossilization to take place is generally too long for the dead organisms to be preserved in mineral form, and therefore fossilization is a rare phenomenon as it is selective and complex. Only in a handful of special contexts – notably xeric or anoxic – is degradation slowed or stopped, allowing a few traces to remain in the sediments over long periods of time, and thus, by diagenesis (see Box 1.1), become fossils.

For a color version of all the figures in this chapter, see www.iste.co.uk/neraudeau/fossilization.zip.

Fossilization,

If we look at present-day ecosystems, it is estimated that two thirds of marine animals have little or no mineralized elements to their bodies (they have no “skeleton”, in the broad sense of the word). Of the remaining third, which is more skeletonized, only a third has a significant likelihood of being fossilized. In other words, a third of a third of the original animal biodiversity – that is only around 10% of creatures that have lived – have the chance of being fossilized.

> *Diagenesis* refers to the set of processes which occur, one after another, in the transformation of sediments into sedimentary rocks. This includes physico-chemical processes (compaction, desiccation, dissolution, precipitation, crystallization, temperature variations, etc.) and biochemical ones (reducing or mineralizing micro-organisms, etc.). Chronologically, a diagenetic sequence contains first the compaction of the sediment, then cementation, and finally (re)crystallization.

Box 1.1. *The concept of diagenesis*

> *Taphonomy* is the discipline which consists of retracing all the processes – biological (putrefaction, carrion feeding, etc.), chemical (degradation, dissolution, mineralization, etc.), sedimentary (burial, filling, etc.), physical (UV exposure, temperature variations, etc.) and mechanical (transport, fragmentation, erosion, etc.) which have occurred between the death of an organism and its final fossilization. *Diagenesis* (see Box 1.1) has an important place in these various steps, combining sedimentary, chemical and physical processes. It should be noted that certain organisms may have been buried alive. In that instance, their taphonomic history began even before the moment of their death.

Box 1.2. *The concept of taphonomy*

Even within this small proportion of marine organisms which can be fossilized, most only give us incomplete fossil records (the skeleton without the soft tissues, an internal cast, a hollow mold), which are sometimes deformed (due to the pressure of sediment, tectonic motion, metamorphism), or indirect indications (tracks, burrows, coprolites, eggs, bite marks) of the species’ anatomy. In the subaerial environment on land, where the burial of the organic remains in sediments is more random than in an aquatic environment (marine or lacustrine), the potential for fossilization is lower still.

1.2. The main processes of fossilization

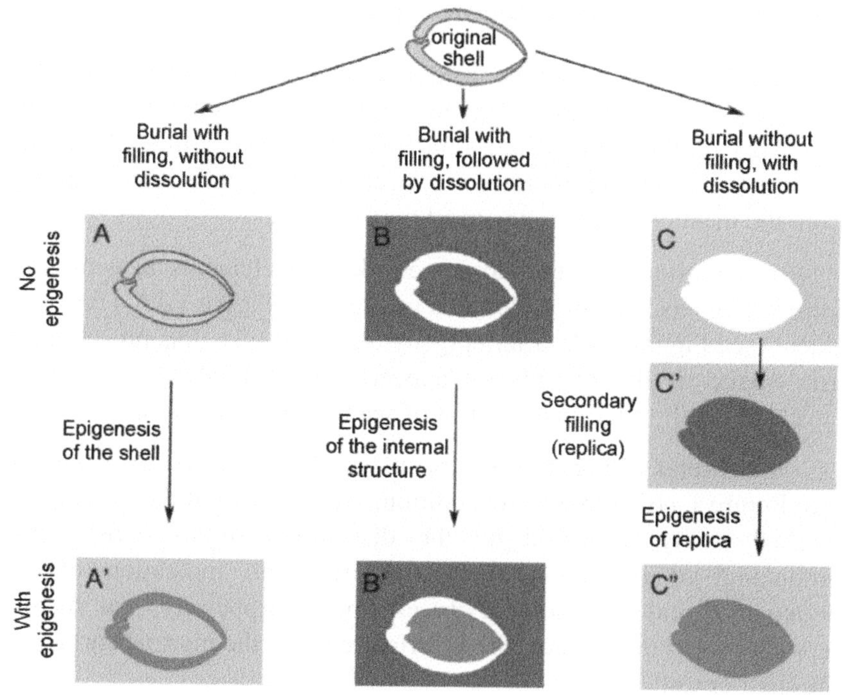

Figure 1.1. *Elementary types of fossilization*

NOTE ON FIGURE 1.1.– *The most common fossilization processes are based on the filling of the empty spaces in the organisms, notably after the decomposition of the soft parts, but also the preservation or dissolution of the mineralized parts. For example, for a bivalve animal (lamellibranch, brachiopod), if burial is accompanied by inter-valve filling, then the shell may be preserved identically (A), or in epigenetic form (A'). If the burial and the internal filling are followed by a dissolution of the valves, the final fossil will be an internal cast (B) made up of the initial intervalve sediment, which may sometimes be epigenetic (B'). In the event that the surrounding sediment does not penetrate the inter-valve space and the valves are dissolved, the final fossil can only be expressed on the walls of the cavity occupying the total volume of the organism (C); in a second step, if a new sediment fills that void, it will form a replica of the organism (C'), and in some cases, that replica may be epigenetic (C'').*

After burial of a skeletonized organism (shell, bone, test), the processes of fossilization depend on the taphonomic history (see Box 1.2) of the mineralized remains. The most commonly found processes can be summarized as three fundamental cases, ranging from the most complete to the most partial:

– preservation of the endo- or exoskeletal elements in a mineral form which has been modified to a greater (see Figure 1.1A') or lesser extent (see Figure 1.1A) after epigeny (see Box 1.3), thus preserving a true image of the organism's appearance, at least of its mineral envelope (shell of mollusks or test of sea urchins, for example);

– quasi-total filling with sediment (see Figure 1.1B) or a mineral (see Figure 1.1B') (epigenetic addition, see Box 1.2) of the organism's internal cavity, followed by the dissolution of its exoskeleton, producing an internal cast, which offers no indication of the organism's real volume and its external morphology, but merely reproduces, on the surface, the inverted relief of the internal surface of the shell or test;

– complete filling, either with sediment (see Figure 1.1C') or a mineral (see Figure 1.1C), of the void left in the rock by the organism's dissolution (see Figure 1.1C), producing a replica which shows the whole of the organism's volume and its external morphology.

On the basis of these three fundamental principles of fossilization, there are numerous combinations which can occur in the wake of the burial of a shell, with or without sedimentary filling, with or without dissolution of the shell, with or without secondary mineralization of the shell or filling of its internal cavities (see Figure 1.2).

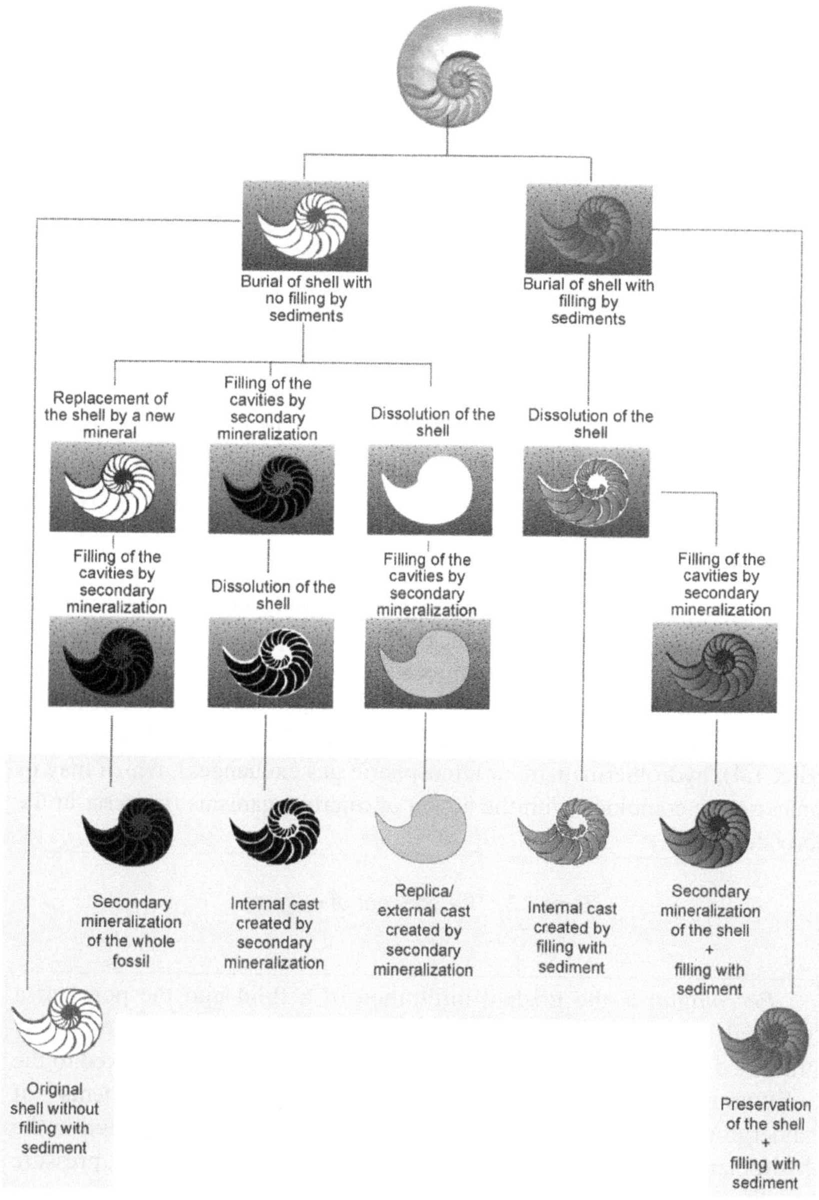

Figure 1.2. *The different variants of modes of fossilization*

NOTE ON FIGURE 1.2.– *Excluding the most elementary modes of fossilization (see Figure 1.1), there are numerous variants in the processes of fossilization. These depend on various combinations between filling or non-filling of the buried remains, whether or not they are dissolved, whether or not they undergo secondary mineralization, and whether a cast is created.*

The primary mineral constituting the skeletal elements of the original organism may be replaced by another mineral by means of isovolumetric substitution (perfect epigeny of the original skeletal elements).

However, this substitution may also occur in a secondary manner, at the level of an internal cast or replica which are formed, during an early stage of diagenesis, in a mineral form which differs from the final form.

> *Epigeny* is a process whereby a mineral body changes its chemical nature without changing its shape. We also speak of epigenetics, or pseudomorphosis. When such a process affects a fossil, it does not alter either its volume or its shape. Epigeny generally occurs as the result of the circulation of a fluid (e.g. percolation of mineralized water, see Box 1.4), hydrothermalism, or atmospheric gas exchanges), which may or may not be combined with the action of micro-organisms (bacteria in the broadest sense).

Box 1.3. *The concept of epigeny*

> *Percolation* is the gradual infiltration of a fluid into the pores of a material. In a geological context, for the most part, sediments are slowly traversed by a liquid (or possibly a gas). This process can be linked to the infiltration of surface water (due to gravity), to the "movements" of underground waters (the water table), but also to injections of fluid in the hydrothermal context, in particular temperature and/or pressure conditions.

Box 1.4. *The concept of percolation*

1.2.1. *Preservation of the endo- or exoskeletal elements*

There is practically no fossil record of vertebrates unless bones, teeth or both are preserved. The organisms' organic phase is typically lost, but their mineral part, composed of calcium phosphate (apatite), is preserved, with a greater or lesser degree of re-crystallization. Teeth are stronger than bones, and are often preserved in sedimentary contexts where the bones have been lost to dissolution. Hence, these are types of fossils which are not exceptional in terms of their nature, and can only be viewed as exceptional in terms of their representativity of such-and-such a species (for example, in view of the rarity of mammal teeth from the Jurassic (–201 to –145 million years ago, Ma), and the Cretaceous (–145 to –66 Ma)). Though bones can less easily be fossilized than teeth, they also represent fairly commonplace fossils, and are exceptional only in particular cases of mineralization or representativity of a taxon.

Fossils of invertebrates with shells (brachiopods, mollusks), carapaces (crustaceans) or test (foraminiferans, sea urchins) constitute as many cases of "ordinary" fossilization as there are taxonomic groups. All these organisms, which have exoskeletons, have good potential for fossilization of their mineral parts, and sometimes the soft tissues are lost during putrefaction. The test of sea urchins is probably the strongest component, with each skeletal element (the plate of the test or radioles) being composed of a very stable calcite from the life of the animal and the remnant during fossilization. However, it sometimes occurs that, during diagenesis, the recrystallization of the test fuses all the plates into a single calcite monocrystal. The aragonite of foraminiferans generally stabilizes as calcite during the course of fossilization, making them one of the most common microfossils, highly present in numerous ancient sediments. Similarly, the most common brachiopods, corresponding to articulate species, usually have a calcitic shell with an external cryptocrystalline layer and a fibrous internal layer, which remains stable during fossilization. Inarticulate brachiopods, whose shell combines calcium phosphate and chitin, are more fragile and less capable of withstanding diagenesis. In the carapace of crustaceans, calcium carbonate and chitin are combined. This latter ingredient, when it is abundant, renders the mineral envelope fragile, meaning it cannot easily be fossilized, and becomes fine and powdery, or even ghostly.

1.2.2. *Formation of replicas or countermolds*

Plate 1.1. *The three types of fossil preservations obtained for the same gastropod mollusk (Pleurotomariidae)*

NOTE ON PLATE 1.1.– *A: limestone replica of the original shell with detailed preservation of the ornamentation of the shell's external surface; B: hollow imprint in surrounding limestone, showing the negative of the relief of the shell's ornamentation; C: internal cast of the shell, with filling of the spire towers in the form of a "rolled-up sausage" in a spiral shape (collection: IGR – Institut de géologie de Rennes; photos: D. Néraudeau). Scale bars: 1 cm.*

Often, organisms with a carbonate shell, especially mollusks, have their mineralized parts composed of aragonite, which is a type of calcium carbonate that is not very resistant to the dangers of diagenesis (thermal or acid "attack").

When that fragile shell is dissolved during diagenesis, its internal void (e.g. between the two valves of a bivalve or in the spire of a gastropod) sometimes remains free from early filling by sediments. The sediment containing the organism then contains a cavity in which the void left by the dissolved shell is found in addition to the void left by the animal's internal cavity. This total cavity has exactly the same volume as the original organism and the sediment which surrounds it, when it hardens, shows the negative of the *mold* of that volume (see Plate 1.1B).

If, subsequently, a new sediment infiltrates and fills that void, it forms a *countermold* of the original organism (see Plate 1.1A). This copy is generally called a *replica*, and usually forms in a carbonated context; it is chalky in nature. Examples of replicas formed in marl or sandstone are also known, but are rarer. The replica may have been created or altered by the percolation (see Box 1.3) of a mineralized fluid and be composed of a homogeneous mineral (for example calcite, opal or pyrite).

1.2.3. *Formation of internal molds*

When, during or shortly after its burial, a shell is filled with sediment, this sedimentary content will become lithified (see Box 1.5) and form a solid mold of the internal space between the valves (bivalves, brachiopods) (see Figure 1.1B) or inside a shell (ammonites, gastropods) or a test (sea urchins).

Following the hardening of that internal mold, the shell surrounding it may be dissolved (for example by acid attack by means of fluid percolation, see Box 1.3) and the internal mold then “floats” in the rock where it was formed, with the thickness of the dissolved shell creating a void between the internal mold and the surrounding rock (see Plate 1.1C). The surface of the internal molds reproduces the inverted relief of the internal surface of the original shell.

For example, in bivalve mollusks, casts of the surfaces where the adductor muscles attach, which are hollow on the internal faces of the shells, correspond to protuberances in the relief on the surface of the internal mold.

In sea urchins, the small holes corresponding to the ambulacral pores in the respiratory areas which pierce the test correspond to small “bumps” in the relief on the internal mold.

Lithification is the successive processes of compaction and cementing by which an initially mobile sediment is transformed into hardened sedimentary rock. Through compaction, the sediment loses the majority of its fluid content, notably liquid, and undergoes packing, losing volume to a greater or lesser extent. The cementing phases correspond to diagenetic crystallizations or recrystallizations (see Boxes 1.1 and 1.2), which can equally affect the particles of the sediment and the porosity between them.

Box 1.5. *The concept of lithification*

1.2.4. *Case study: geode formation*

Geode formation (see Box 1.6) is a special case, which corresponds to the crystallization of a highly mineralized fluid having percolated into certain cavities of the shells. Largely, this occurs when the organisms' internal cavities (total or partial) are not filled with sediment when they are buried (see Figure 1.3).

Unlike replicas or recrystallized internal molds, geodes do not fill all of the internal cavities of the shells. They merely coat the walls with crystals with centripetal growth, where the points generally converge towards a central space, which remains hollow (see Figure 1.3B'). The mineral nature of the crystals is determined by the chemical composition of the percolating fluid.

For example, in chalky terrain it is fairly common for the percolation of a carbonated fluid to cause calcite geodes to form in the cavities which it passes through. Similarly, in a context rich in silica (notably decalcification alterites), the percolation of a silicated fluid will form quartz geodes.

Besides these relatively frequent cases, numerous mineral species can form geodes in the original cavities or the dissolution cavities of fossil organisms.

It should be noted that independently of the geodes crystallizing in the internal cavities of fossils, crystalline outgrowths can sometimes develop on the outer surface of the shells or bones, when they are buried in a highly porous sediment or without direct contact with a portion of the fossilized remains (see Figure 1.3C').

A *geode* is a cavity lined with crystals within a rock (be it sedimentary, igneous or metamorphic). The cavity in question is sometimes that of a shell from a mollusk or brachiopod, a test from a sea urchin, or any other hollow part of a skeletonized organism. From germs (impurities, points of erosion or dissolution, etc.) on the internal surfaces of the cavities, crystals grow as a result of the percolation water or hydrothermal fluids which bring mineral compounds that make up the mineral species in the geode.

Box 1.6. *The concept of a geode*

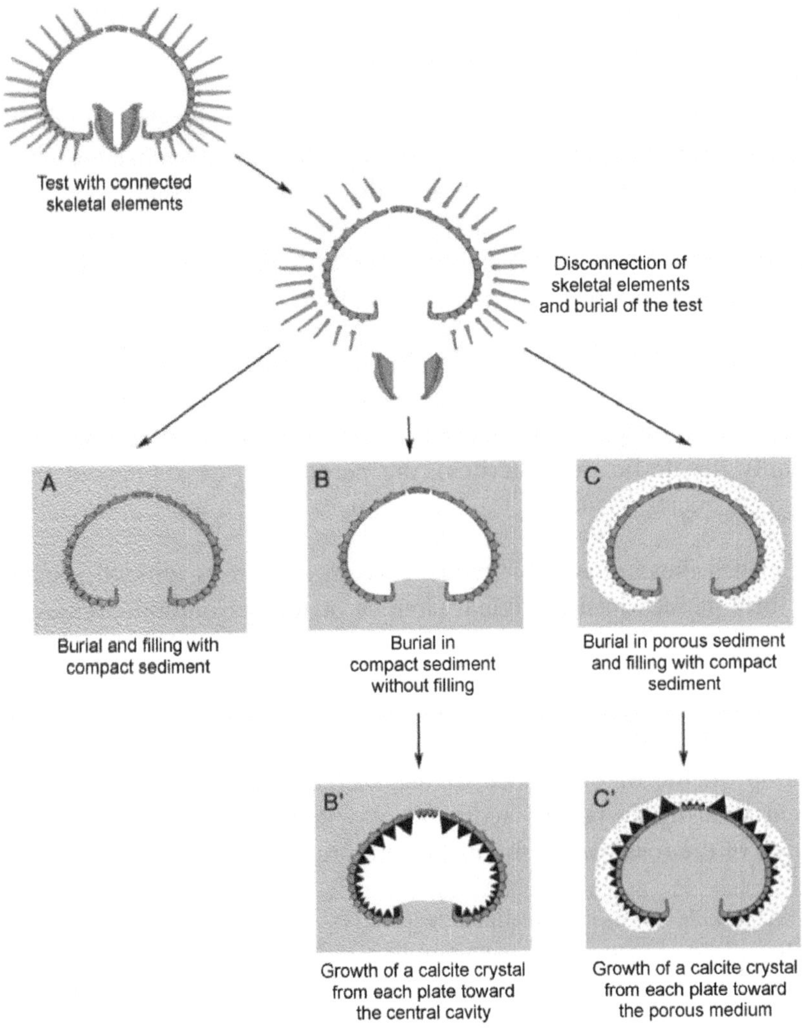

Figure 1.3. *Geodes and crystalline crusting*

NOTE ON FIGURE 1.3– *The burial of a shell or a test and its filling generally takes place in parallel with compaction of the sediment, which leaves no gaps either at the surface or inside the shelled remains (A). However, it sometimes happens that the internal cavity – for example, the inside of a sea urchin's test – is not filled with sediment (B). If a mineralized fluid (notably containing calcium carbonate or silica) percolates into the surrounding sediment, it can reach that internal cavity and crystallize there on the walls, forming a geode (calcite, quartz or other minerals) (B'). It may be that an organism's internal void is completely filled with compact sediment, but a substantial amount of porosity remains in the sediment that is in contact with the external surface of the shell, the test or even the bone (C). The percolation of a mineralized fluid into the surrounding sediment may lead to crystallization in that porous space, with crystals encrusting the outside of the organism (C').*

1.3. The concept of a Lagerstätten

In rare cases, certain processes of mineralization, with favorable contexts and facies, can lead to truly exceptional cases of preservation. These remarkable cases occur either sporadically, affecting only a few specimens at a local scale, or more generally across a fossiliferous site, a geological formation or a sedimentary layer.

In the occasional cases, it is a matter of exceptional preservation in the strict sense, yielding unexpected, spectacular, esthetically beautiful and particularly rare fossils. Above all, this relates to the phenomena of original mineralization/crystallization, fossilizations preserving the integrity and anatomical connections of the specimens, or the preservation of trophic situations involving two or more species (prey/predator, host/parasite associations, etc.).

In the more generalized cases, exceptional preservation relates to the vast majority of the groups and individuals associated in the same paleontological deposit. This type of deposit exhibiting exceptional preservation was dubbed a "Lagerstätte" (plural: Lagerstätten) by Adolf Seilacher in 1970. This term means a "storage place", referring to the fossil content of a sedimentary deposit. There are two types which are differentiated by the exceptionality either of the quality of the fossils or of their quantity.

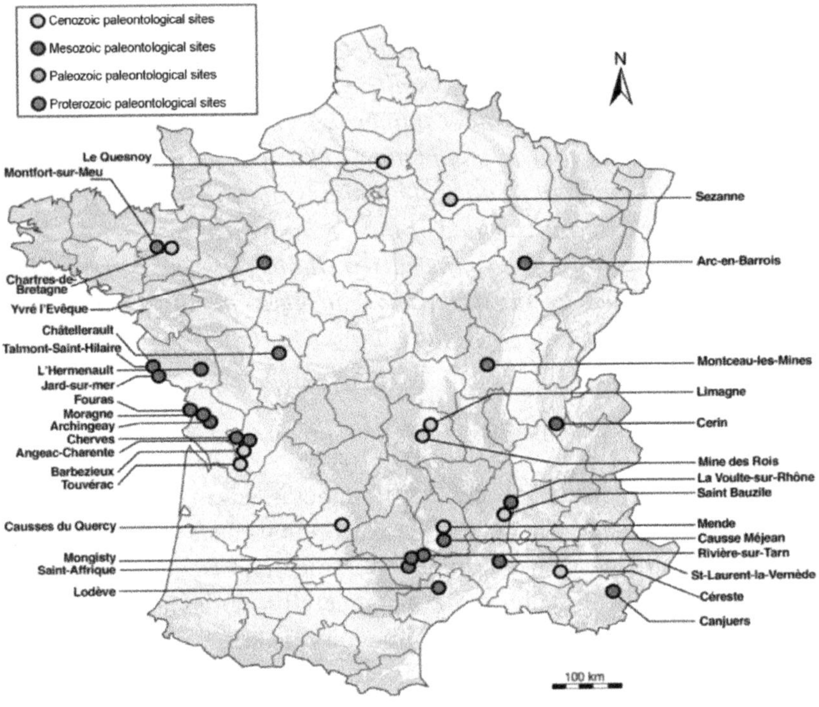

Figure 1.4. *Map of fossil deposits in France*

NOTE ON FIGURE 1.4.– *Map showing the location of paleontological deposits in France which are mentioned in this book, differentiated by the eon to which they belong: Proterozoic (pink), Paleozoic (green), Mesozoic (blue), or Cenozoic (yellow).*

The name "Konservat-Lagerstätten" is applied to facies which have an exceptional quality of preservation of the fossils, with almost complete preservation of anatomies, partial anatomic connections, 3D fossilizations, mineralization of soft tissues and variable combinations of these different remarkable cases.

The name "Konzentrat-Lagerstätten" applies to sites where an exceptional quantity of fossils are found, be they bones or tracks of vertebrates (the most usual cases), casts or cuticles of plants, or shells or tests of invertebrates. In such a context of remarkable accumulation, the fossilized organisms are not necessarily exceptional in terms of the quality of their fossilization, and they may well be less than complete or highly fragmented, with anatomical connections or dispersed elements.

This book gives an overview of France's most remarkable mineralizations, facies and deposits in terms of original or exceptional fossilization, also touching on a number of fossils from outside of France to illustrate cases that are not found in the national territory (see Figure 1.4).

2

Diversity of Types of Mineralization

2.1. Silicification

There are many different processes involved in silicification, which, despite different geological histories, may lead to very similar exceptional fossils. Silicified fossils may result from what is known as primary silicification, roughly contemporaneous with the sediment deposit and the death of the organisms in question, or from secondary silicification, which occurs millions of years after the sedimentation of the organisms. In certain cases, fossilization may be the result of processes that combine prompt and delayed silicification.

Early silicification can occur by the involvement of *endogenous* silica, which is present in a highly localized or highly diffuse form in the architecture of an organism. For example, the spicules of a sponge – once they are dissolved – can supply a portion of the amorphous silica which then precipitates, silicifying the mass of the sponge's body (other possible examples can be found in diatoms and

For a color version of all the figures in this chapter, see www.iste.co.uk/neraudeau/fossilization.zip.

radiolaria). Relatively early and *exogenous* silicification may occur by the reuse of amorphous silica from siliceous organisms, such as sponges, but precipitating in the remains of other organisms associated with the sponges, such as mollusks and echinoderms.

Delayed silicification tends to use exogenous silica, involving the amorphous silica carried by the circulation of fluids (percolation, alteration, leaching, hydrothermalism) from one sedimentary medium to another. Thus, fossiliferous chalk beds covered by (stratigraphic superposition) or adjoining (lenticular systems) sandy formations subject to leaching may see the silicification of fossils contained therein.

2.1.1. *Primary silicification*

In the case of primary silicification, we can distinguish between cases resulting from the hydrothermal impregnation of organisms near to silica-rich fluid sources, and cases resulting from diagenetic or synsedimentary recrystallization of the remains of non-siliceous organisms.

2.1.1.1. *Hydrothermal or volcanic silicification of land-dwelling organisms*

Example of the Rhynie chert

The region of Rhynie in Scotland was the site of a great deal of volcanic activity in the late Devonian (around –396 Ma), which led to the formation of hot springs, generating silica-rich fluids (Channing and Edwards 2009). Vents and natural events (such as geysers) brought, or ejected, these “siliceous gels” to the surface, where they covered various organisms and impregnated them with silica, even silicifying their cellular ultrastructures (Kerp 2018). The most remarkable silicification was of plant matter – notably, the earliest forms of subvascularized terrestrial plants (tracheophytes), such as *Rhynia*, which were originally named in reference to the name of the fossiliferous deposit site (see Plates 2.1A and 2.1B). The paleontological assemblage also includes bryophytes, lycophytes such as *Asteroxylon*, fungi such as ascomycetes and freshwater algae such as charophytes. In addition to these plant remains, the site contains small arthropods and nematodes, remarkably preserved in 3D (Dunlop and Garwood 2018). The paleoenvironment, which was the backdrop to this exceptional fossilization, aside from its volcanic context, was a marsh. The Rhynie cherts reflect the fertile soils for varied plants that were part of the process of plant colonization of the dry land.

Plate 2.1. *Fossils silicified in a volcanic context*

NOTE ON PLATE 2.1.– *A and B: polished sections of the Rhynie cherts, from the Devonian of Scotland. A: general view of a chert with a multitude of sections of axes of* Aglaophyton*; B: wide view of a piece of chert with sections (basically transversal) of axes of* Aglaophyton*; C: general view of carbonized wood impregnated with infiltration of hyalite opal in a volcanic context, Le Fau, Cantal; D and E: terrestrial gastropods of the genus* Helix, *mineralized with orbicular lussatite, coming from Tertiary alterites from Mine des Rois, Dallet, Puy-de-Dôme (collections: A and B: Nigel Trawin, University of Aberdeen; C: Maison de la minéralogie du Massif central, Antignac; D: Néraudeau-IGR; E: J.-C. Le Sager; photos: A and B: C. Strullu; C and D: D. Néraudeau; E: M. Guillo). Scale bars: 1 cm.*

Example of Neogene wood with hyalite opal

In Cantal, and more broadly in Auvergne, volcanic activity contributed to the fossilization and mineralization of terrestrial organisms and, more specifically, plant remains.

Thus, in addition to silicification with lussatite (see section 2.1.2.2), wood was preserved in the late Neogene (around –11.5 to –2.5 Ma) in carbonaceous form (carbonized to a greater or lesser degree) by volcanic ash or lava flows, and sometimes originally mineralized by the percolation of silica-rich fluids, which crystallized in the form of hyalite opal (the vitreous form with the formula SiO_2nH_2O).

This exceptional process is notably seen in the Fau sector, in Pays de Salers, and in other sites in Puy-de-Dôme (Flörke et al. 1973; Lynne and Campbell 2004; Barailler 2023) (see Plate 2.1C). The impregnation of the wood with opal makes for a striking contrast between the black color of the carbonized wood and the luminous transparency of the hyalite opal.

2.1.1.2. *Early diagenetic silicification of marine organisms*

The Anglo-Paris Basin contains a thick Cretaceous series of chalk deposits, mainly laid down between the end of the Cenomanian (around 95 Ma) and the start of the Maastrichtian (around 70 Ma).

This chalk corresponds to a relatively deep marine deposit (50 to 200 m depending on the sites and periods), in the context of a substrate favorable to the development of diverse benthic organisms such as echinoderms (for example the sea urchins *Echinocorys* and *Micraster*), bivalve mollusks (for example *Inoceramus*) and sponges (for example *Siphonia*).

These organisms, as the result of the degradation of their siliceous "gridlike" skeleton and, above all, their siliceous spines – known as "spicules" – produced huge quantities of microscopic silica debris which accumulated in the chalk as the strata were laid down.

The proliferation of burrowing organisms during certain periods of the deposition remobilized this silica with the effect of concentrating it. Limnivorous burrowers, as they passed through, ingested the sediment, consumed and thus removed the organic matter, and naturally excreted mud with an altered pH and chemical composition, poorer in organic carbonate particles and richer in siliceous particles, which the burrowing organism was unable to assimilate. This results in precipitations of silica where the mineral was present in excess concentrations – i.e. in the burrows created by the organisms.

Silicified burrows within the chalk therefore form remarkable beds (see Figure 2.1A): black in color, winding and labyrinthine, networks of flint, as can be seen in the cliffs of Étretat (Normandy, north-western France) (see Plate 2.2A).

In addition to silicified burrows, the flint-bearing chalk beds contain the remains of invertebrates and vertebrates that have been partially or completely transformed into silica, particularly near the burrows, such as sponges, echinoids and the teeth of selachians (see Plate 2.2B to 2.2E).

These primary flints, and the fossils they contain, can be found in alterites, after the dislocation and dissolution of the limestone gangue (see Figure 2.1C).

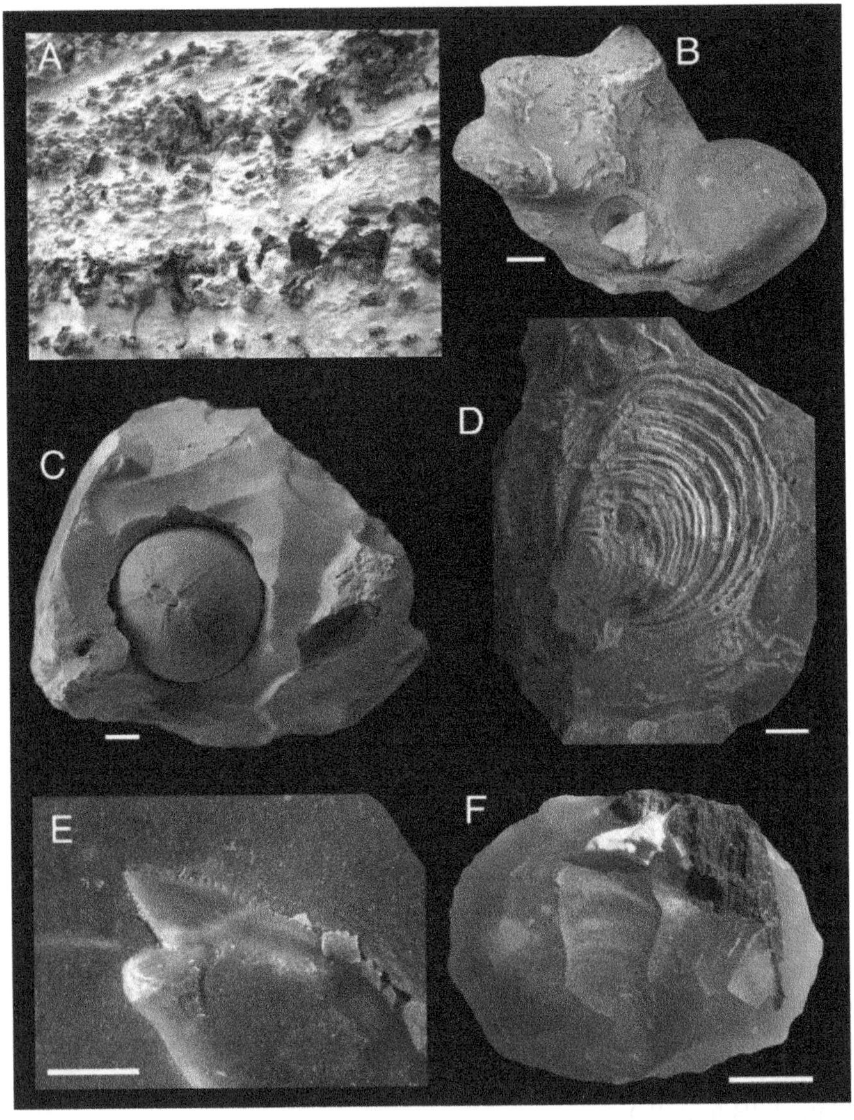

Plate 2.2. *Fossiliferous primary flint*

NOTE ON PLATE 2.2.– *A: chalk beds rich in winding black flints, interspersed between white chalk layers with lower flint content, in the cliffs of Étretat, Seine-Maritime, dating from the Turonian; the* *winding form of the flints reflects the bioturbation of the levels at which they were formed; B: twisted nodule of flint which has partially silicified and encapsulated a sea urchin (*Echinocorys*) in the process of nodulization; C: flint nodule which has totally silicified and encapsulated a sea urchin (*Echinocorys*), whose calcareous test was dissolved by decalcification; D to F: Turonian flint fossils from the Cher region (central France): D: large bivalve (*Inoceramus*) totally silicified in a lump of flint; E: shark tooth (*Squalicorax*) encapsulated in a flint nodule, the large cuspid of its crown having remained apatite; F: lignitic wood caught within a flint nodule, without having totally lost its carbonaceous nature (collections: B and C: IGR; D to F: K. Chartier; photos: A: taken in situ by E. Lasseur; B to E: D. Néraudeau; F: K. Chartier). Scale bars: 1 cm.*

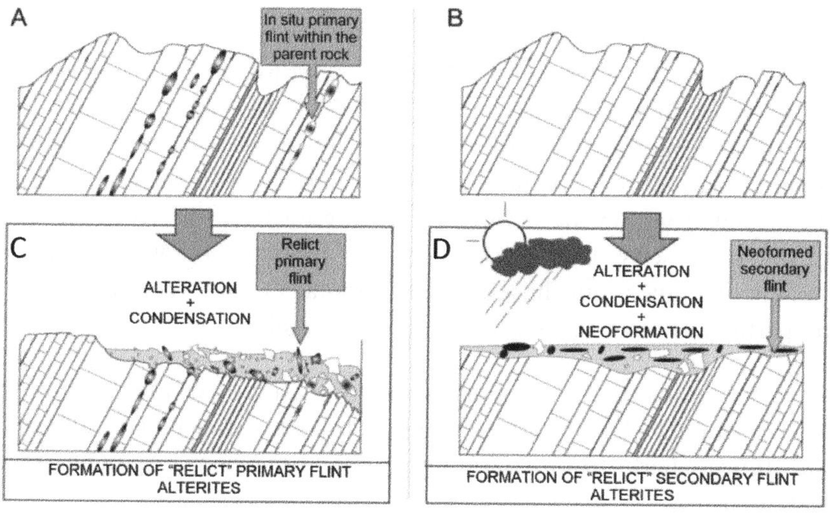

Figure 2.1. *Flint formation*

NOTE ON FIGURE 2.1.– *Flint formation can occur in various geological contexts. Synsedimentary flints, known as "primary flints", are formed at the time of the initial deposition of the sediment containing the flints – generally chalky limestone – and are encased in those chalky beds (A). The alteration of the surface of these flint-containing beds dissolves the surrounding carbonate rock, but preserves the flints in the alterites in the form of relict primary nodules (C). However, a series of limestone beds which initially contained no nodules of flint (D) can lead to their formation by means of the alteration of the exposed parts (D). This alteration produces decalcified clay in which flint re-forms, generally a long time after the deposition of the original limestone. Thus, in western Europe, the formation of such secondary flints, meaning neoformed, generally took place at the end of the Paleogene, to the detriment of Cretaceous limestone. In these neoformed flint formations, there is often a great deal of time-averaging, meaning a collection of fossils from successive geological strata, which are concentrated in the same layers of flint (D) (adapted from Moreau et al. (2016a)).*

Example of marine flints from the Turonian in Cher

The chalky limestone of the Turonian (–94 to –90 Ma) is often rich in layers of flint, present in the form of twisted nodules which are indicative of bioturbations of the marine substrate. All elements of the benthic marine (or floating terrestrial) fauna are preserved, either encapsulated within the nodules, or emerging from their surface to a greater or lesser extent. These Turonian flints are known in various regions of France, but those examples found in Cher (central France) are particularly rich in fossils and original fossil species.

The species that are most frequently silicified are echinoids, but we can also find asterids, sponges, bivalves (see Plate 2.2D) and sometimes shark teeth (see Plate 2.2E) or wood fragments (see Plate 2.2F).

In these latter cases, silicification has proceeded in the direction of the teeth or fossilized wood without having totally epigenized them, and a significant amount of apatite remains in the teeth, or carbonaceous material in the wood, despite the generalized silicification of their gangue.

Sometimes, this partial silicification of fossils encapsulated in flint nodules in certain deposits from the Turonian in Charentes (western France), even goes as far as the preservation of non-epigenized grains of amber (a polymerized resin), within the siliceous mass (Néraudeau 2015).

2.1.1.3. *Early diagenetic silicifications of terrestrial plants*

Permineralizations (see Box 2.1) are characterized by the three-dimensional preservation of the plant's tissues and cells.

In certain cases, the cuticle and all tissues are preserved. This type of preservation is the most informative for paleobotanists.

In extremely rare cases, the preservation is so extraordinary that even the cell nuclei and chromosomes may be preserved (Bomfleur et al. 2014).

Permineralized tissues are characterized by micro-crystallization in the intercellular spaces, in the cells or along the cell walls. By definition, the nature of the precipitated minerals may vary (for example calcite, silica, pyrite).

In cases where the cell walls themselves are mineralized, we no longer speak of permineralization but of "petrification" (Martín-Closas and Gomez 2004). Also, permineralization must not be confused with "authigenic cementation", where mineral precipitation is limited to the periphery of the tissues, affecting neither the tissues themselves nor their cells.

> *Permineralization* is a process of mineralization that takes place when water penetrates the porosity of an organism that is being fossilized. On the microscopic or nanometric scale, this water percolation, when it is mineralized, leads the empty spaces to be filled by the precipitation of the minerals in solution.

Box 2.1. *The concept of permineralization*

Examples of plant-bearing flints from the Middle Jurassic, found in Haute-Marne (eastern France)

Bathonian limestone (–171 to –168 Ma) from the Arc-en-Barrois sector was partly deposited in a shallow marine environment with a significant input from the land. This rock yields a particular type of flints, known as cherts, in the form of nodules shaped like "pillows" or "loaves" (see Plate 2.3A), containing plant debris (wood, conifer branches, pteridosperm fronds, bennettitalean pinnules) (Le Couls et al. 2015; Le Couls 2017) (see Plate 2.3B, 2.3C and 2.3D).

These plants are preserved as permineralizations in three dimensions with all the tissues remaining, from the outermost to the innermost (see Plate 2.3E). The chalk and cherts in Arc-en-Barrois contain sponge spicules in abundance, sometimes on the surface of the plant matter. The spicules are likely to have been a major source of silica and played a major role in the silicification of the organic remains.

As has been demonstrated by in vitro experiments, plants (particularly wood) are excellent "host structures" for silica (Laroche et al. 1989). The high affinity of some of their organic compounds (for example lignins) with silica can lead to the rapid silicification of certain tissues (Laroche et al. 1989). In such a context, in Arc-en-Barrois, the plant debris may have served as nuclei during the formation of siliceous nodules, which may have begun during sedimentation or during early diagenesis.

It should be noted that we tend to speak of flints when speaking of the relatively translucent and shiny siliceous nodules from the Upper Cretaceous, and of cherts when referring to the matte and opaque siliceous outcrops from the Middle or Upper Jurassic. Cherts still contain a little calcium carbonate whereas true flint, which is more vitreous, no longer contains any.

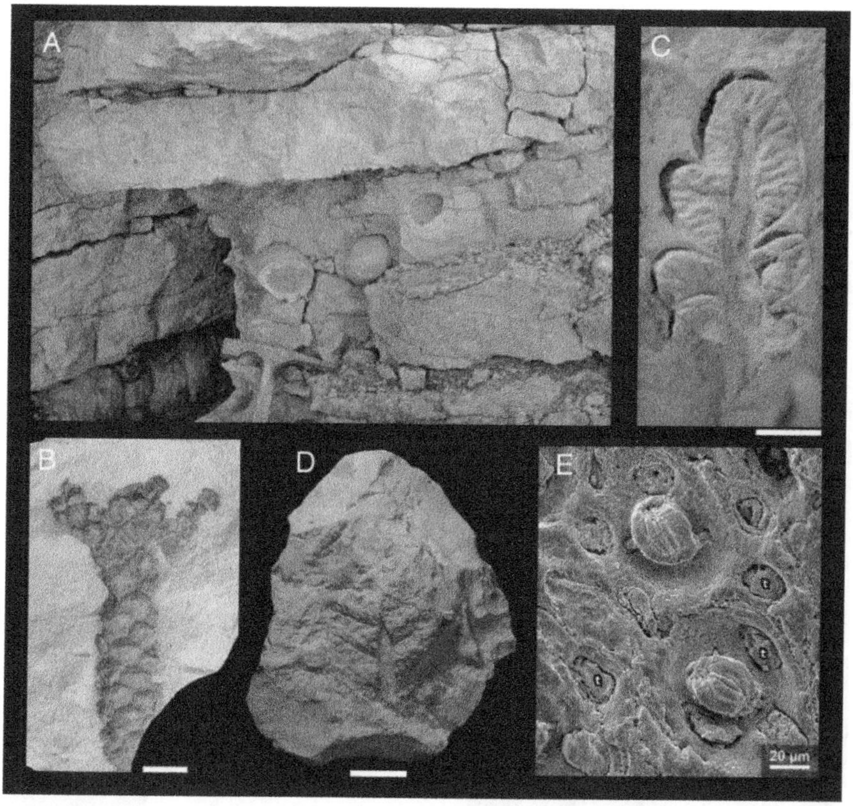

Plate 2.3. *Silicification of plants in cherts*

NOTE ON PLATE 2.3.– *A: ovoid chert blocks in situ, in the Bathonian limestone of Arc-en-Barrois, Côte-d'Or (western France); B: leafy branch of a conifer* Brachyphyllum desnoyersi *from the Callovian deposit in Laignes, Côte-d'Or; C to E: a section of a frond (D) from the fern* Cycadopteris moretiana *from the Bathonian, Arc-en-Barrois, Côte-d'Or, with a detailed view of the end part of a pinna (C) and scanning electron microscope view of silicified stomatal apparatus (E) (collection: B to E: Le Couls-IGR; photos: M. Le Couls). Scale bars: B = 5 mm; C = 2 mm; D = 1 cm; E = 20 µm.*

2.1.2. *Secondary silicification*

Plate 2.4. *Spicules and orbicules of silica entering into the processes of secondary silicification, with examples of fossils from the Upper Cretaceous, found in Charente-Maritime (western France)*

NOTE ON PLATE 2.4.– *A: "empty phantoms" of sponge spicules in an opalescent matrix, formed by the reprecipitation of the silica from those spicules when they dissolved (Cenomanian limestone, silicified during the Eocene, found in Archingeay, Charente-Maritime, western France); B: aggregate formed of debris of sponge spicules, within a flint, viewed with synchrotron microtomography (Cenomanian limestone, silicified during the Eocene, found in Archingeay); C: a fragment of a leaf limb (lighter-colored area), within a flint, covered with sponge spicules (Cenomanian limestone, partly silicified during the Eocene, found in Moragne, Charente-Maritime, western France); D and E: fossilized oyster shell (D: internal surface; E: external surface), part of which retains the lamellar carbonate structure (white), while other parts are in the process of silicification by orbicules of calcedonious silica (pink) (Campanian fossils, partly silicified during the Holocene, found in Barzan, Charente-Maritime, western France); F: fossilized oyster shell in the process of centrifugal silicification by orbicules of calcedonious silica (pink) (Campanian fossils, partly silicified during the Holocene, found in Talmont, Charente-Maritime, western France) (collection Néraudeau-IGR; photos: A, C to F: D. Néraudeau; B: J.-D. Moreau-ESRF). Scale bars: A and B = 0.2 mm; C = 5 mm; D to F = 1 cm.*

In the case of secondary silicification, we can distinguish: (1) events caused by a generalized decalcification of carbonate rocks, in combination with a precipitation of silica (see Figure 2.1D); and (2) more isolated events affecting a few calcareous fossils gradually recrystallizing to form orbicular or rough chalcedony. The chalcedony orbicules develop gradually, centrifugally, beginning at a microscopic scale, and then growing concentrically. Originally dispersed throughout the carbonate mass, they ultimately grow together, finally coming to completely substitute the calcium carbonate (see Plate 2.4D to 2.4F). On the other hand, rough chalcedony tends to fill a void left in the sediment by the total or partial diagenetic dissolution of a shell, or to fill the internal cavity of an organism (between two valves of lamellibranchs, in the spire of a gastropod, in the test of a sea urchin) which was not filled with sediment when it was buried post-mortem. Thus, it has a tendency to form geodes (see Plate 2.5D). Chalcedony is a type of silica that has a fibrous microstructure, less hydrated than opal, which only contains between 0.5 and 2% water (Lacroix 1910; Devouard and Bornet 2015). It sometimes encrusts the cavities in sedimentary rocks and thus the cavities in the fossilized shells (various mollusks, echinoids) where it has been brought by the percolation of a silica-rich fluid (see Box 1.3).

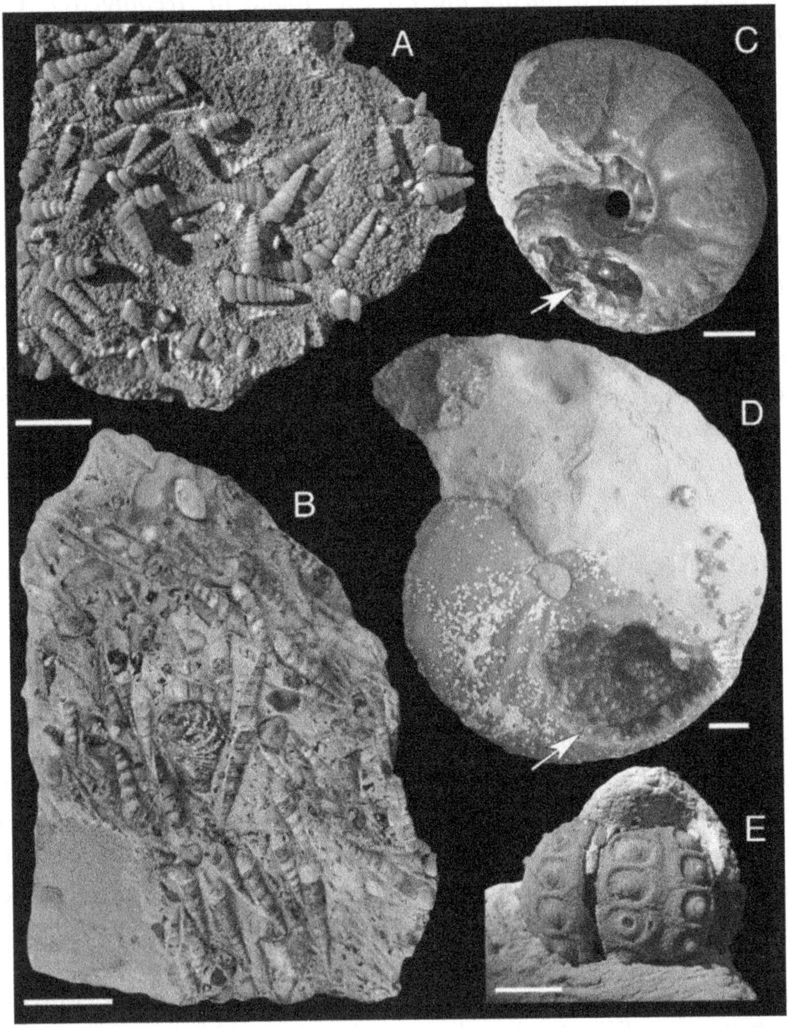

Plate 2.5. *Various chalcedony fossils*

NOTE ON PLATE 2.5.– *A: gastropods* Potamides lapidum *with translucent chalcedony (Lutetian, found in Maule in the Paris region); B: marine gastropods with opalescent chalcedony (bluish appearance due to alteration), found in Belgium; C: ammonite* Callihoplites auritus *(from the alterites from the Albian, Seine-Maritime, north-western France) with translucent chalcedony, with rough chalcedony formations in the internal cavities (indicated by the arrow); D: indeterminate ammonite (found in alterites from the Turonian in Río de Oro, Morocco) with translucent chalcedony, having rough chalcedony formations in the internal cavities (indicated by the arrow); E: sea urchin of the Cidaridae family, recrystallized with translucent chalcedony (Turonian, found in Cher) (collections: A: É. Dépré; B: B. Hogrel; C: IGR; D: J.-C. Le Sager; E: K. Chartier; photos: A, C, E: D. Néraudeau; B: B. Hogrel; D: M. Guillo). Scale bars: A, C, E = 1 cm; B = 4 cm; D = 5 mm.*

It should be noted that the appearance of fossil replicas in secondary chalcedony may change over time. It can change from a translucent crystalline appearance (see Plate 2.5A), when the mineral is not altered, to a bluish-white, milky appearance when the surface film of the chalcedony is altered by dehydration (see Plate 2.5B). This process of alteration is comparable to the patina which forms on the surface of prehistoric knapped flint that has been exposed to atmospheric or pedological attack for a long time (Cayeux 1930).

2.1.2.1. *Mass Paleogene silicification as a result of climate*

Example of Eocene–Oligocene flints containing Cretaceous plants

The Paleogene (–66 to –23 Ma), notably the Eocene (–56 to –34 Ma), was a period with warm, humid episodes accompanied by processes of lateritization which leached carbonates from the exposed marine Mesozoic layers. This phenomenon has been widely documented in western and northern France, and southern Britain (Gallois 2009). It leads to the formation of alterites, also known as “flint clays”, which correspond to time-averaged strata, in which we sometimes find fossils of widely varying age (Moreau et al. 2019a) (see Figure 2.1D). When the Cretaceous limestone contains a high proportion of silica, notably in the form of sponge spicules, the processes of dissolution/recrystallization have affected these siliceous micro-elements, sometimes leaving microvoids where they were dissolved, and forming an opalescent matrix of the re-precipitated silica (see Plates 2.4A to 2.4C).

At the macroscopic scale, this amorphous silica has become concentrated in the form of nodules, and sometimes substituted the original carbonates in certain fossils. For example, in decalcified clays, the replacement of calcite by silica has led to internal casts of sea urchins (originally made of carbonate) being preserved in siliceous form – in rarer cases, with the test also being preserved.

However, the most spectacular example is the replacement of carbonaceous tissues of fossilized plants, originally preserved in paralic marl or lagoon limestone, by silica. It not only replaces the carbonates from the gangue of plant fossils, but also the residual organic part of fossilized plants, even down to a cellular level (Moreau et al. 2015, 2021). Such cases of siliceous permineralization (see Box 2.1) have been widely observed in flints containing Cretaceous plants (from the Cenomanian and Turonian) in western France. Given the exceptional preservation of the tissues of certain conifers from the Cretaceous in the Charentes and Vienne regions (western and central France), it is highly likely that early silicification, during the sedimentation of the Cretaceous deposits or during an early phase of diagenesis (Moreau et al. 2017a), linked to the abundance of spicules (see Plate 2.4C), preceded the secondary silicification in the late Paleogene (Upper Eocene to Oligocene). Thus, we see that silicification can take place in multiple stages (Moreau et al. 2017a).

This silicification of plants can apply to wood (see Plates 2.6A to 2.6D), leafy axes and reproductive structures, such as cones (see Plate 2.6E), as well as to perforations and burrows made by xylophage invertebrates (insects in an aerial environment, teredo in an aquatic environment) (see Plates 2.6B to 2.6C) (Robin et al. 2018). In such cases, we see exceptional preservation of the silicified plant microstructures, right down to the cellular level (see Plate 2.6D).

Plate 2.6. *Silicified conifer wood, xylophages and cone*

NOTE ON PLATE 2.6.– *A: section of a branch (*Agathoxylon gardoniense*) from the Cenomanian, found in Moragne, Charente-Maritime (western France); B and C: silicified wood with perforations made by aquatic xylophages (*Teredo*), mineralized in opalescent chalcedony, from the Cenomanian deposit of Moragne, Charente-Maritime (western France); D: SEM view of a silicified wood fragment from the Cenomanian deposit of Moragne, showing the fibrous microstructure and the arrangement of cells; E: the strobile of silicified Bennettitale (*Clathropodium trigeri*) from the Turonian, found in Yvré-L'Évêque, Sarthe (central France) (collections: A to D: Néraudeau-IGR; E: Musée Vert, Le Mans; photos: A to C: D. Néraudeau; D: S. Beurel; E: N. Morel). Scale bars: A to C = 1 cm; D = 0.2 mm; E = 5 cm.*

Example of Paleogene burrstone

Burrstones are siliceous rocks typical of the early Tertiary Era (Eocene-Oligocene), formed by rapid silicification, on a geological timescale, of lacustrine marls or limestone (Freytet 1984; Ménillet 1987). The irregularity of the decalcification often leads to rough, irregular, hollow rocks, with multiple alveoli; for this reason, they are often used as grindstones.

The chemical transformation of the original carbonate rocks was the result of a dry terrestrial climate which caused the lakes to dry up and the silica to be remobilized. As siliceous burrstones come from lacustrine carbonate deposits, they preserve the fauna of mollusks, with the silicified shells of freshwater gastropods, such as hydrobia, lymnaea, ramshorn snails and *Viviparus* (Alimen 1947). Also, and above all, they have the peculiarity of containing a large amount of various remains of charophyte algae, whether it is their fructifications (called "oogonia" or "gyrogonites") (see Plates 2.7C and 2.7D), their axes (see Plates 2.7E and 2.7F), or their root networks (see Plates 2.7A and 2.7B). Burrstones represent a unique case of exceptional preservation of charophyte accumulations, not limited to their oogonia. It should be noted that, in very exceptional circumstances, insects (Fric 1884) and foraminifera (Alimen 1947) have also been discovered in such deposits.

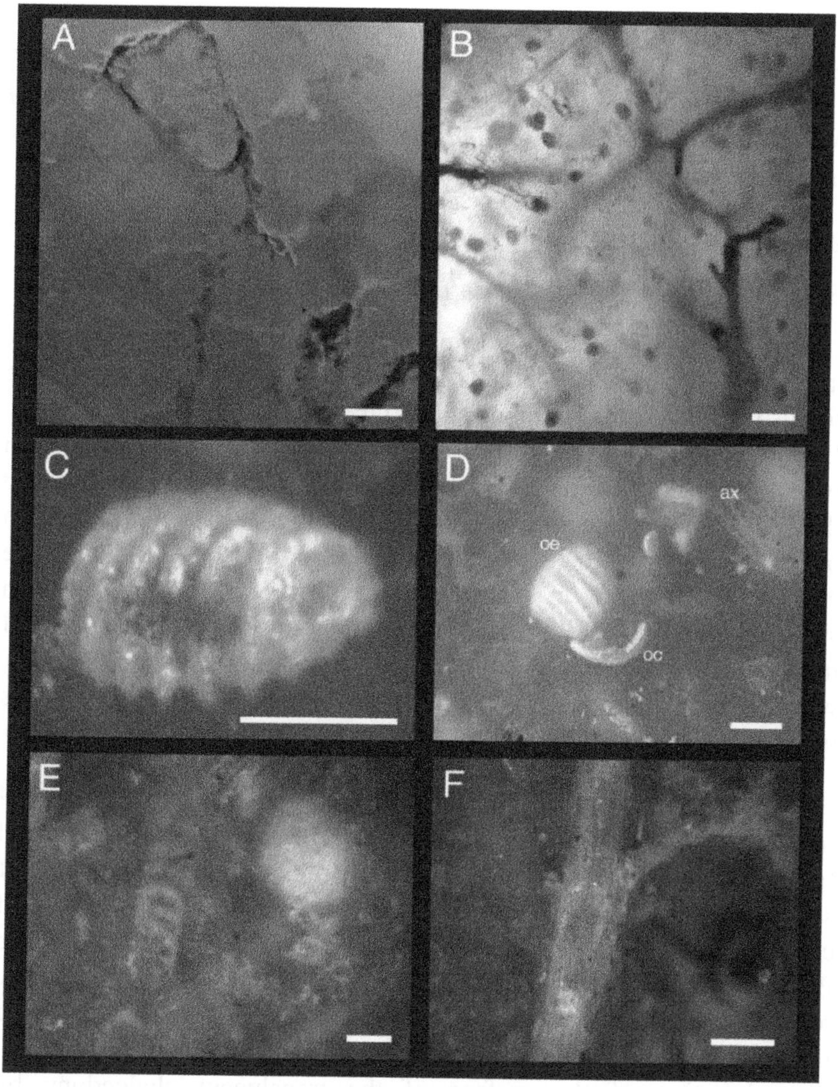

Plate 2.7. *Silicified fossils in burrstones – different parts of charophyte algae preserved by silicification in translucent burrstones from the Oligocene, found in Malaigre, Charente-Maritime, western France*

NOTE ON PLATE 2.7.– *A: root network; B: root network associated with numerous oogonia; C: close-up of an oogone; D: intact oogone (io) and broken oogone (bo), associated with portions of stems (ax); E: close-up of a section of a stem, left, and in the background, right, an oogone; F: close up of a section of a stem (collection Néraudeau-IGR; photos: A and B: D. Néraudeau; C to F: R.-A. Benoit). Scale bars: A and B = 2 mm; C to F = 0.5 mm.*

2.1.2.2. Replicas and internal casts formed of silicates

Example of mollusks and sea urchins in chalcedony

The ability of silicated fluids to infiltrate and precipitate in fractures, large voids or pores sometimes leads to partially or fully crystalline fossils (or formed of amorphous silica) included in non-siliceous rocks. During an initial diagenetic phase, if a fossil has been dissolved, then within the rock it has left a cavity that is identical in both volume and detail to the organism's initial morphology. In a second mineralization phase, the silica was then able to crystallize in that void during a single episode of filling, or in successive stages of development. We then speak of the addition of silica. Chalcedony, which is a fibrous form of microcrystalline silica, sometimes fills the internal cavities of fossils, where it forms bumpy geodes on the walls, notably inside ammonites (see Plates 2.5C and 2.5D).

However, it is also possible for silica to grow by gradual replacement of the mineral forming the replica of the original fossil (generally carbonates), in which case it is more a process of substitution. Thus, microcrystalline silica, such as translucent chalcedony, can replace the carbonate replica of a gastropod (see Plates 2.5A and 2.5B) or sea urchin (see Plate 2.5E). Such chalcedony replicas of mollusks, notably Cerithiidae, are frequently found in the Eocene deposits in the western Sahara (Dahkla region, Morocco). The alteration of the surface of the translucent chalcedony by dehydration can cause an opaque bluish patina that gives it an opalescent appearance (see Plate 2.5B).

Regardless of the previous process involved, the "final" fossil found by paleontologists millions of years later is a replica of the original organism, either totally or partially made of silicates with a crystalline or glassy appearance.

Example of snails preserved in lussatite from Limagne (central France)

Volcanic activity in the Massif Central began around the late Oligocene or Early Miocene (around –23 Ma), in a partly boggy terrestrial landscape. The eruptions caused explosion craters which were sometimes filled with water to form lakes, with the lava flows from the eruptions surrounding the lakes and covering the ground around them. These layers of pyroclastites (eruption residue) were formed of a mixture, known as "peperite", of basalt and carbonate debris from the destroyed substratum. During a lull, the peperites were recolonized at points by vegetation and the associated fauna – notably pulmonated snails of the genus *Helix*. New eruptions produced ash which, in turn, covered the populations of snails. Hydrothermal silica stemming from leaching from these volcanic deposits then filled the voids in the shells to a greater or lesser extent, and also the tubular structures of the associated plants. This local silica is particularly esthetically attractive because it is a bluish chalcedony called "lussatite" (silica containing 3 to 8% water), whose color and shine stand out from the surrounding pyroclastic rock, which is brownish and dull. It should be noted that the development of lussatite in snail shells took place by means of a process of growth and agglomeration of silica orbicules (see Plate 2.1D and 2.1E). The successive layers of silica built up around a nucleus, generally a small fragment of peperite or limestone, and thus formed layered orbicular structures (Devouard and Bornet 2015).

2.1.2.3. *Substitution of carbonates by silicates*

Example of Cretaceous shells silicified during the Quaternary

Many fossils of mollusks and echinoderms have become detached from coastal cliffs of clay and clay-rich limestone formed in the Cretaceous due to wave action. Fairly quickly (on the scale of a human life), the portion of the shells that comes into contact with the air through erosion is covered by small circular siliceous structures in concentric areas: chalcedony orbicules (Lacroix 1910). In the space of a few years or decades, a shell, which was essentially carbonated when it was still encased in the clay gangue, becomes siliceous in its entirety by means of a juxtaposition of hundreds of orbicules – small structures with concentric growth (see Plates 2.4D to 2.4F). Though it is frequently found and easy to observe, little is known about this phenomenon in terms of its dynamic and causality. However, we can point firstly to the role of pH and CO_2 pressure, which change when the shell emerges from its limestone gangue and is exposed to atmospheric conditions and, secondly, to the abundance of silica in particulate form (sponge spicules, see Plate 2.4C), in the limestone – silica which will come to percolate with the groundwater which has dissolved it. However, the orbicules had already begun to develop inside the carbonate shell, and the dissolution of the carbonate revealed them whilst also facilitating their growth.

2.1.2.4. *Fossils with quartz geodes*

Silica is present in abundance in numerous original sedimentary facies (fluvial sands, coastal sandstone, sponge-rich chalk, etc.) or ones resulting from decalcification (flint clays, siderolithic silcrete, etc.). A portion of that silica may be dissolved by percolation or groundwater. When the resulting silicated fluid, as it filters through the rock, encounters empty spaces (fractures in the rock, cavities in the fossils), it crystallizes there forming quartz geodes. The crystalline outgrowths then partially fill the internal cavities of the fossils but can also sometimes encrust the external walls. It is possible to observe fossils with geodes for almost all varieties of quartz, hyaline or colored. It may just as well be varieties of translucent quartz tinted red or yellow by the presence or iron (see Plate 2.9A and 2.9B), tinted blue-gray or black by the presence of manganese (see Plate 2.9C), or lightly tinted pink or purple by the same chemical elements. Geodes of pale amethyst (purple quartz) have been found in the cavities of ammonite or nautiloid shells dating from the Bajocian, found in Normandy north-western France (–171 to –168 Ma), from the Bathonian, found in Nièvre, central France (–168 to –165 Ma), from the Callovian, found in Vosges, eastern France (–165 to

–161 Ma), and from the Oxfordian (–161 to –155 Ma) found in Ardennes, north-eastern France (see Plate 2.9D) (Lebocey 2019). A wide variety of fossils made of citrine (yellow-orange quartz) have been found in China, though citrine can also form replicas, rather than merely automorphous crystals.

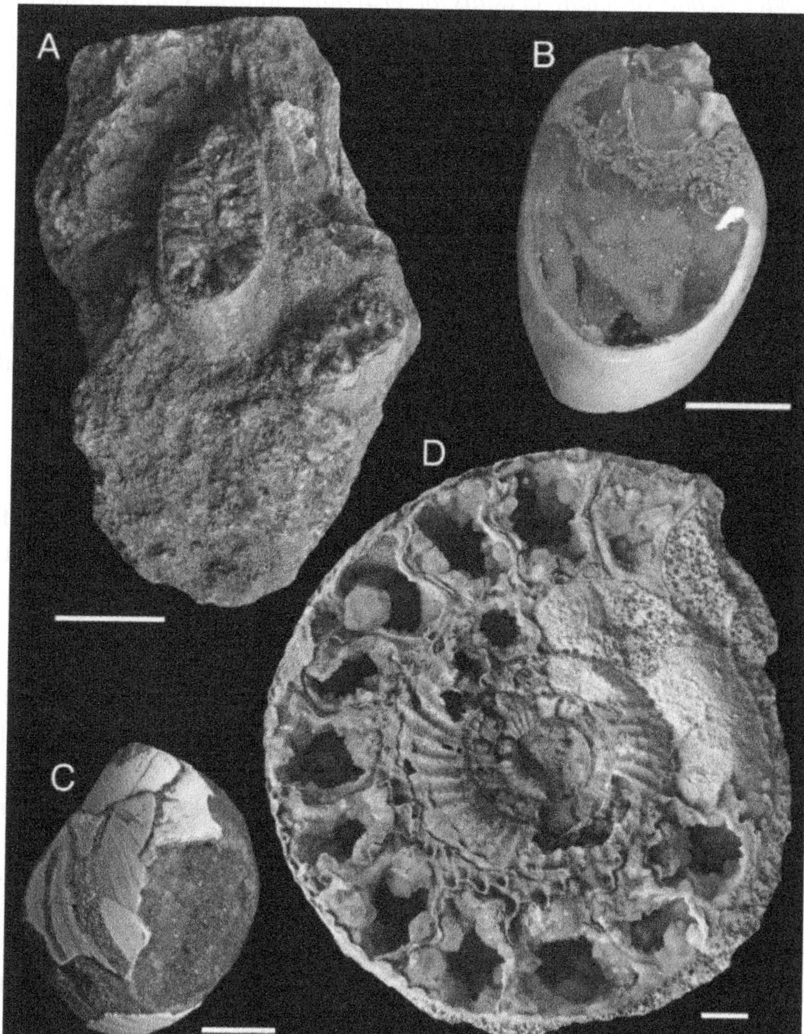

Plate 2.8. *Fossils with crystallization of quartz geodes*

NOTE ON PLATE 2.8.– *A: rudist with its cavities filled with geodes of ferruginous quartz and hematite, dating from the Cenomanian, found in Moragne, Charente-Maritime, western France; B: gastropod with a geode seeded with small crystals of yellowish quartz, dating from the Albian, found in the Aravalli Mountains, India; C: gastropod with a compact geode of small crystals of gray-blue quartz, dating from the Mio-Pliocene, found in Tjörnes, northern Iceland; D: ammonite with chambers transformed into geodes of large crystals of slightly pinkish or purplish quartz (amethyst), from the Upper Bajocian, found in the Verson region, Calvados, north-western France (collections: A and B: IGR; C: É. Dépré; D: P. Lacroix; photos: D. Néraudeau). Scale bars: 1 cm.*

2.1.2.5. *Fossils made of gems and fine siliceous stones*

In addition to the previous cases, there are particularly attractive examples of silicate replicas, when the silicates involved are precious or semiprecious. Such is the case, notably, of fossils in iridescent opal from Australia, and fossils in emerald from Colombia.

Example of fossils in precious opal found in Australia

The opalized marine fossils from the Coober Pedy deposit, discovered in Australia in 1915, were formed by siliceous pseudomorphs, during the Oligocene (–34 to –23 Ma), of carbonate fossils originally formed during the Lower Cretaceous (around –140 Ma), then dissolved during the Upper Cretaceous. In the Oligocene, a fluid, enriched with silica through processes of surface alteration, percolated into the fossiliferous rocks and filled the void left by the dissolution of fossils which were originally replicas or internal casts (mainly bivalves, gastropods and belemnites) (Robertson and Scott 1990; Thiry et al. 2006) (see Plates 2.9C and 2.9D). Comparable replicas of terrestrial fossils (plants, vertebrates) have also been found on another Australian site: Lightning Ridge (see Plate 2.9A and 2.9B) (Mustoe and Smith 2023). Opal ($SiO_2.nH_2O$) is a form of silica containing a variable amount of water (between 2 and 20%) in its mineral microstructure. Precious opal, which displays iridescent colors, generally contains 6–10% water. Such precious opal is composed of microspheres of amorphous silica (diameter about 2 to 4×10^{-6} mm) with regular diameter and arrangement. Light is diffracted by these regular layers of microspheres, and thus split up into the different colors of the spectrum (rainbow effect). By comparison, common opal, which is not iridescent, is formed of microspheres of silica with non-uniform diameters and arrangements, which do not lend themselves to this spectacularly colorful diffraction of light (Smallwood et al. 2008).

Plate 2.9. *Fossils formed of opal. Different types of fossils from Australia exhibiting pseudomorph into opal*

NOTE ON PLATE 2.9.– *A: cone, Lightning Ridge; B: tortoise vertebra, Lightning Ridge; C: gastropod, Coober Pedy; D: belemnites against gangue of alterites, Coober Pedy (collections: A and B: Australian Opal Centre; C and D: J.-C. Le Sager; photos: A and B: J. Brammail; C and D: M. Guillo). Scale bars: A to C = 0.5 cm; D = 1 cm.*

Example of emerald fossils found in Colombia

The most exceptional of the "precious" silicate fossils are indubitably the emerald ($Be_3Al_2Si_6O_{18}$) gastropods found in Gachalá, Colombia (Vuillet et al. 2002; Pignatelli et al. 2022).

In a particular context of temperature (around 300°C), hydrothermal fluids enriched with beryl, drawn from the surrounding rock, percolated, during the Late Cretaceous, into the porosity of a black limestone argillite dating from the Berriasian (–126 to –121 Ma, Lower Cretaceous), rich in organic matter (2%) and pyritous fossils – particularly ammonites (Branquet et al. 1999). These hydrothermal fluids have sometimes reached replicas of gastropods stemming from an initial phase of fossilization (after the dissolution of the aragonite shell) and, by metasomatism (see Box 2.2), replaced them either totally or partially with green beryl (see Plate 2.10A and 2.10B). This is a dual process of dissolution/substitution of the original shells and replicas which leads to their pseudomorph into emerald.

It should be noted that pyrite is associated in varying quantities with beryl in fossils, and that black shales also contain silicate fossils made of albite ($NaAlSi_3O_8$) (Vuillet et al. 2002).

Metasomatism is a metamorphic process linked to the circulation of hydrothermal fluids, in which one mineral is replaced by another in specific conditions of temperature and/or pressure. The original mineral is replaced by a new mineral composed of chemical elements brought in by the hydrothermal fluid.

Box 2.2. *The concept of metasomatism*

Fossils formed of green garnet (iron and calcium silicates), unearthed in Madagascar, correspond to organisms from the Upper Triassic (–237 to –201 Ma; polyparies and gastropods), transformed by hydrothermal metasomatism (Lebrun 2015).

In all these cases, the fossils were initially formed during the Mesozoic (–252 to –66 Ma), then transformed in the Cenozoic (–66 Ma to today) by processes of addition or substitution which introduce opal, beryl or garnet in place of the original mineral replicas.

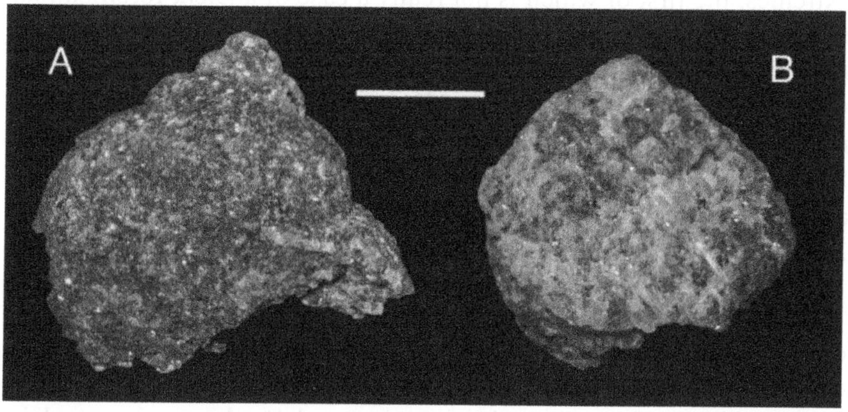

Plate 2.10. *Emerald fossils. A and B: gastropods fossilized with "pseudomorphs" in emerald (metasomatism process), from the Matécaña mine, Gachalá sector, Colombia (collection: J.-C. Le Sager; photos: M. Guillo). Scale bar: 0.5 cm*

2.2. Carbonization of plants and compressions of cuticles

Carbonization processes apply mainly to plants. However, they may be the result of different conditions, sometimes linked to incomplete combustion or simply exposure to heat in the context of a fire, sometimes far removed from the influence of any fire, in an oxygen-poor aquatic environment.

2.2.1. *Carbonification due to anoxia*

In the absence of free oxygen ("gaseous" oxygen), anaerobic bacteria "consume" the oxygen present in the organic molecules and thus increase the concentration of carbon. In this way, plant remains which have not putrefied due to lack of oxygen are transformed into various types of carbonaceous rocks, from peat, over a short span of geological time (early diagenesis) and containing only around 50% carbon, to anthracite, over a long span of geological time, containing over 90% (Laveine 2007; Baudin et al. 2017). Between these two extremes, the first of which is not really a case of ancient fossilization, and the second of which does not preserve the anatomical details of the plants, there are various intermediary stages, including lignite which corresponds to around 65–75% carbon, but preserves the plants, notably wood, right down to the fine details of their cellular structure. Peat, lignite and anthracite are on a gradient of increasing temperature and pressure associated with a greater burial depth (Baudin et al. 2017).

In an aquatic environment, wood that has floated away from land on the waterways, during floods or storms, and accumulated in the coastal context or in a paralic basin, has been shielded from putrefaction by the salt in the seawater and by the marine mud in which it was buried. In the paralic basins from the Mesozoic or Cenozoic, we most often find fossilized wood in lignite form, showing preservation of all the characteristic histological structures of the original wood: tracheids, cross-field pits or growth rings.

2.2.2. *Carbonization in connection with forest fires*

The exposure of plants to fire and their combustion in an oxygen-poor environment leads to the production of coal/charcoal (Scott 2000; Baudin et al. 2017). Such carbonization due to fire dispels the more volatile compounds (water, hydrogen, oxygen), and enriches the material with carbon. Carbonization through fire produces mega- and mesoremains which have increased resistance to diagenetic compaction. Thus, plant remains after carbonization due to fire often exhibit three-dimensional structures that have undergone only little to moderate compaction, and are very black in appearance.

In addition, wood charcoal has the property of being extremely brittle, which explains why certain geological formations yield a multitude of small fragments while larger specimens are rare (Friis et al. 2011). It is interesting to note that this type of preservation is found extremely frequently in fossilized wood, but also applies to very delicate, complex and diminutive structures such as reproductive organs (Friis et al. 2011). Carbonization due to fire sometimes provides remarkable histological information.

Note also that the degree of preservation of tissues depends on the temperature to which they were exposed. Based on experiments conducted with contemporary coniferous wood, Scott (2000) showed that beyond a certain threshold (280 to 320°C), we see homogenization of the cell walls. Beyond 370°C, the cell walls begin to crack and break, with the damage increasing up to the terminal point of 600°C.

In the case of fossilized coal, Friis et al. (2011) noted that the plant's outermost envelope (the cuticle) is often indistinguishable, being more or less homogenous with the rest of the structure. The fossilized wood typical of these carbonization contexts is charcoal, which is somewhat powdery and blackens the fingers when one touches it, but still exhibits the very finest of microstructures. We also find "vitrified" fossilized burnt wood, presenting conchoidal fractures, called jet. In the first case (charcoal), diagnostic anatomical characteristics are present, and we can identify the wood from which it formed (and the plant family to which it belonged); in the case of jet, it is not possible to do so.

Fossilized carbonized wood, originally set alight by lava flows, may also have been preserved in a volcanic context. Such is the case, for example, with the carbonized tree trunks found in the Permo-Triassic basalt of Aït-Aadel in Morocco. The wood (*Araucarioxylon?*) is not only carbonized but has been subject to the circulation of copper-rich fluids (hydrothermalism, leaching from cupriferous rock), which have given it tinges of blue or green due to micro-encrusting of copper-based minerals such as bornite (Cu_5FeS_4), malachite ($Cu_2CO_3(OH)_2$) and azurite ($Cu_3(CO_3)_2(OH)_2$) (Moussu 1959; Jenny 1988). In addition, above (see section 2.1.1.1), we have mentioned wood that has been carbonized in a volcanic context with encrustations of hyalite opal.

2.2.3. *Biogeochemical carbonization (also known as "diagenetic carbonization")*

So-called "ligniteous" deposits are frequently observed in the fossil record. Often, they contain plant remains which appear to be carbonized, but which have not undergone combustion. They may have undergone "biogeochemical carbonization", also known as "diagenetic carbonization" (Martín-Closas and Gomez 2004). Unlike carbonization due to combustion, which is almost instantaneous (at least on a geological time scale), geochemical carbonization is associated with the gradual carbonization of a plant's original structures, as diagenesis increases in intensity (growing over time, greater depth of burial, greater temperature and pressure). The strongest components of the plant are affected last (for example lignins, cutanes). The most advanced stages of diagenetic carbonization are responsible for the world's large coal deposits (for example deposits from the Carboniferous, –359 to –299 million years). Lignite is one of the earliest stages in coal formation. By comparison with a plant carbonized by fire, a plant that undergoes geochemical carbonization will be more affected by diagenetic compaction. Lignites show a varying degree of compaction, which is a function of the rate of carbonization and the intensity of diagenesis. Whilst coal is quite abundant in strata dating from the Carboniferous, lignite is more frequently found in the Mesozoic – in particular, the Cretaceous – and in the Cenozoic. The ligniteous microstructures of the wood are generally fairly well preserved in these deposits, and the plant families to which they belong can often be identified (mainly Araucariaceae and Cheirolepidiaceae from the mid-Cretaceous).

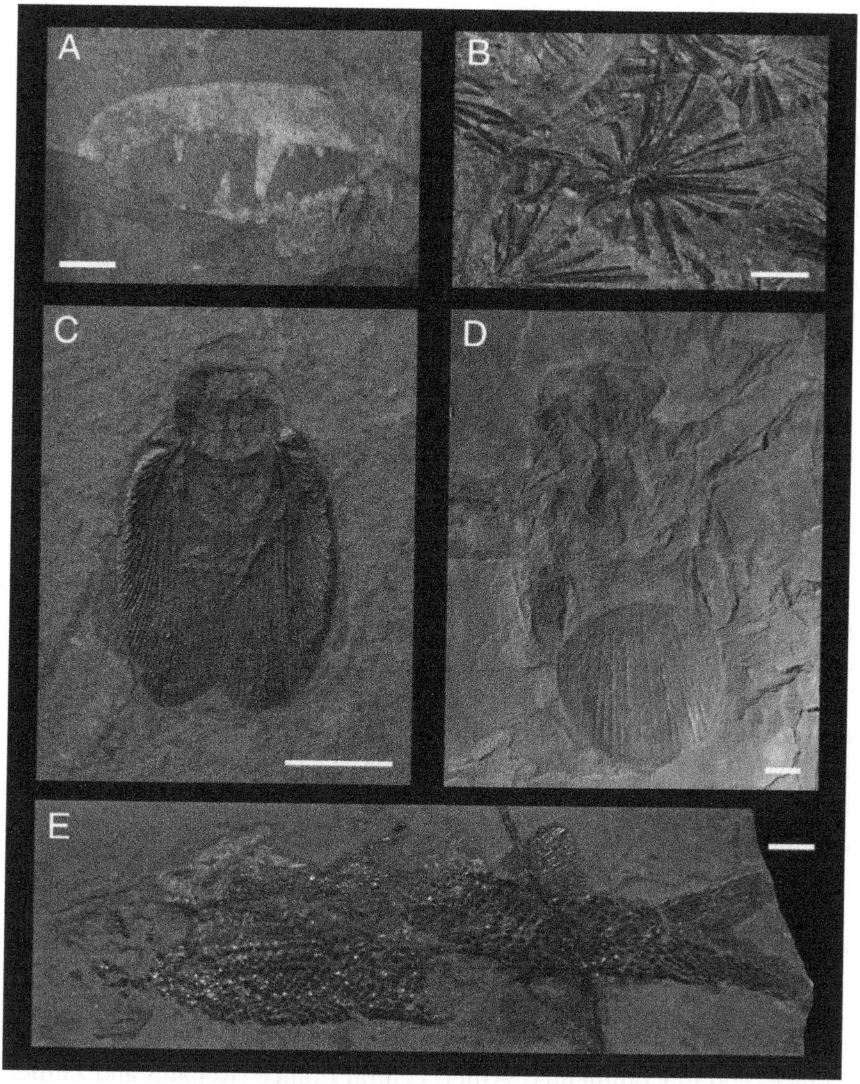

Plate 2.11. *Fossils in coal or ampelitic facies: organisms preserved in high-carbon deposits*

NOTE ON PLATE 2.11.– *A: eurypterid pincer, ampelite shale from the Silurian in Saint-Denis-d'Orques, Sarthe (north-western France);*

B: Sphenophyte Annularia longifolia, *from the Carboniferous, found in Saint-Pierre-La-Cour, Mayenne (north-western France); C: blattoptera insect* Etoblattina gaudryi, *from the Stephanian, Carboniferous found in Commentry, Allier, (central France); D: head and portion of the torso of the eurypterid* Megarachne servinei, *from the Carboniferous coal deposits of Bajo de Veliz, Argentina; E: actinopterygian fish* Aeduella blainvillei, *dating from the Autunian in the Permian, from Autun, Saône-et-Loire, (central France) (collections: A: Gendry-IGR; B to E: IGR; photos: D. Néraudeau). Scale bars: 1 cm.*

Note that the same layer may resemble charcoal, jet and ligniteous wood, particularly in coastal facies (paralic basins) where wood debris, which may or may not have been burnt, has been transported and brought together by estuarine currents.

Limnic, marshy or lacustrine basins, far from the influences of the sea, were the environments in which coal tended to form and are characteristic of the Paleozoic coal deposits rich in fossilized plants. The plant species in these deposits, locally, are abundant and highly diverse, in the form of molds and countermolds for fern fronds and the leafy axes of horsetails (see Plate 2.11B), or in the form of three-dimensional replicas, for the vertical growth axes (pseudo-trunk) or subterranean growth (pseudo-roots and rhizomes).

In addition to plants, coal facies may also contain fossils of arthropods and fish (see Plates 2.11C to 2.11E).

Coal series are generally associated with the Carboniferous deposits, such as in Saint-Pierre-La-Cour, Mayenne, north-western France, and in Commentry, Allier, central France (Renault and Zeiller 1890; Béthoux and Nel 2003). However, there are also comparable deposits dating from the Lower Permian, such as those from the Autunian (the Carboniferous–Permian transition), which alternate coal layers and bituminous shale (Broutin et al. 1999; Bercovici and Broutin 2008).

The curious case of Silurian ampelites

In the Silurian (–444 to –419 Ma), fine clay sediments, rich in organic matter, were deposited in relatively deep seas with little oxygen. These anoxic marine deposits were then buried to varying depths and metamorphized during the Hercynian orogeny and thus compacted, heated and laminated.

However, their high levels of pyrite, linked to their original anoxic environment, helped to preserve organisms of varying sizes, the most remarkable of which are small graptolites (colonial plankton forms closely related to hemichordates) and giant eurypterids (arthropods known as "sea scorpions") (see Plate 2.11A and 2.11D). In these deposits, the fossils are usually observable as compressions of white residue contrasting with the blackish carbonated surrounding rock. The rocks associated with this very particular type of carbonaceous fossiliferous facies are known as "ampelitic shale", or ampelites, with very high carbon levels. The Silurian ampelite on the Armorican Massif in western France is one of the best-known examples due to the abundance and diversity of graptolites (for example *Monograptus*, *Cyrtograptus*) found there (Philippot 1950; Piçarra et al. 2002), along with some eurypterid remains (see Plate 2.11A).

2.2.4. *Plant compressions*

"Compressions" are the most common form of plant preservation in the fossil record (Schopf 1975; Taylor et al. 2009) (see Figure 2.2). Generally speaking, they are preserved in lignite and clay, less frequently in limestone and exceptionally, in sandstone (Taylor et al. 2009). They include a multitude of types of fossils, from wood to foliar remains or reproductive structures. The significant compression of certain plant remains can largely be explained by the diagenetic compaction of the surrounding rock (see Figure 2.2). The compression of the fossils also depends on the nature of the buried remains. For example, wood charcoal produced during a forest fire will be much more difficult to compact than a delicate structure like a flower. The quality of preservation of the compressions depends not only on the taphonomic processes (for example mechanical degradation, fungal degradation or decomposition linked to phytophagous or xylophagous insects, the speed of burial, the degree of oxygenation of the environment, the nature of the sediment), but also the intensity and the nature of the diagenetic processes/factors (for example pressure/ dissolution, temperature, depth, percolation).

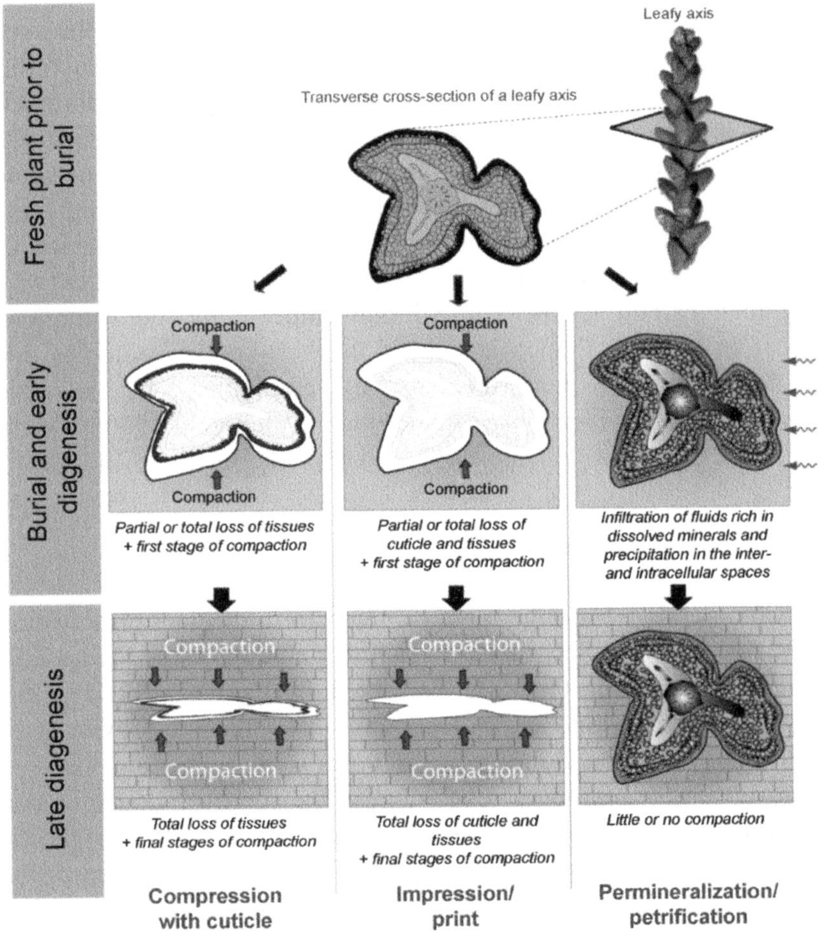

Figure 2.2. *Types of fossilization of foliar plant remains*

NOTE ON FIGURE 2.2.– *Diagrams illustrating the different stages in the formation of the three main types of plant preservation in foliar remains (case of a conifer branch).*

Often, the compressions are limited to the cuticle of the plant (see Plates 2.12A and 2.12B). The cuticle is the plant's outermost layer: it plays a protective role and limits gaseous and aqueous exchanges between the plant's external and internal environments. The cuticle is composed of cutanes, which give it a great deal of resistance to fossilization and cutins, which are hydrophobic polymers (with 16 or 18 carbon atoms). In numerous cases, the compressions may only include the cuticle as the tissues of the plant have not been preserved. Though superficial, the cuticular envelope provides remarkable details on the stomatal apparatus and the epidermal cells (see Plate 2.12F). It also sometimes happens that the cuticle is only preserved in the form of shreds which are mere millimeters or even micrometers in size (see Plate 2.12D). When they are completely preserved, the cuticles can be directly removed from the rocky substrate with a knife or spatula (see Plate 2.12A). In certain exceptional cases, these "fossil leaves", when freshly collected, are still supple, and once they are removed from the substrate, their transparency reveals their nervation in the finest of detail (see Plate 2.12B).

Example of Cretaceous flowers, found in Gard, south-eastern France

Cretaceous flowers are very sparse in Europe (Friss et al. 2011) and similarly rare in France. They provide crucial data for our knowledge of angiosperms during their rapid and early diversification, which took place at the intersection between the Lower Cretaceous and the Upper Cretaceous. Flowers and inflorescences have been discovered in the Cenomanian deposits (–100 to –94 Ma) in Gard (south-eastern France) (Moreau et al. 2016b). These reproductive structures are preserved in the form of brownish compressions (with little compaction), and exhibit astonishing preservation of all parts of the perianth (sepals/petals), the androecium (stamens), and the gynoecium (ovaries), pollen grains that can still be observed in situ by synchrotron X-ray microtomography (see Figure 2.3). Studies have shown the presence of flowers and fruits attributable to Platanaceae and Lauraceae, with anatomical features preserved in never-before-seen detail.

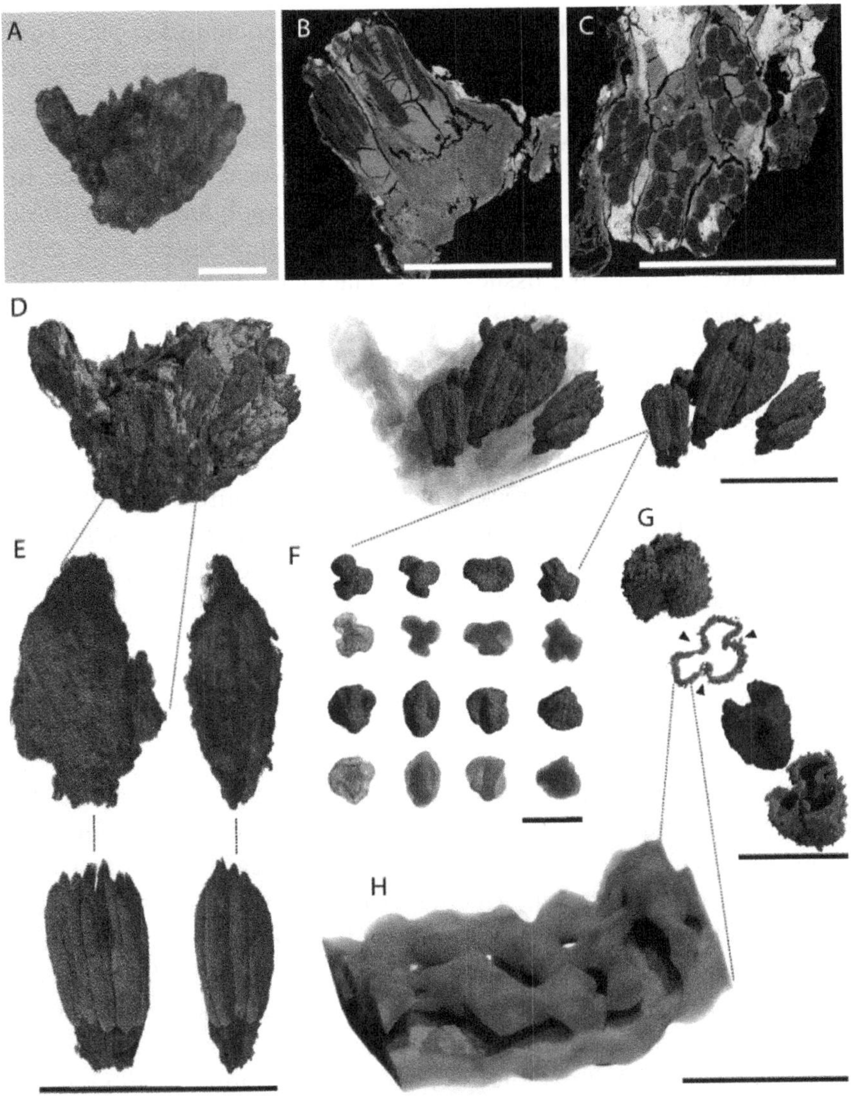

Figure 2.3. *Flowers from the Cenomanian, found in Gard. Synchrotron micro- and nanotomography images*

NOTE ON FIGURE 2.3.– *Inflorescence with male flowers from the Cenomanian at Saint-Laurent-la-Vernède (Gard, south-eastern France), captured by synchrotron phase-contrast microtomography (ESRF). A: fragment of an inflorescence; B and C: virtual cross-sections, longitudinal (B) and transversal (C), of the same specimen; D: 3D reconstructions of the fragment of inflorescence (D) and of a flower (E) showing the perianth (purple), staminal filaments (green) and pollen sacs (orange); F and G: grains of pollen contained in the pollen sacs; H: detailed view of the external wall of a pollen grain, the exine (modified from Moreau et al. 2014). Scale bars: A to E = 1 mm; F = 10 μm; G =10 μm; H = 1.5 μm.*

2.2.5. *Imprints and casts*

An imprint is characterized by the external molding of a plant on a rocky surface (limestone, marl, clay, etc.) and the practical absence of any organic matter (Schopf 1975) (see Plates 2.12C to 2.12E and Figures 2.5C to 2.5E). These are prints and counterprints.

In certain cases, the imprints are created by the destruction/alteration of a compression (Schopf 1975). Though they are prints, when the sediment is very fine (for example clay or marl), certain details, such as the different levels of nervation of the leaves, or the petals of flowers, may be very clearly imprinted on the surface of the rock.

Exceptionally, the imprints can provide information concerning the morphology and dimensions of epidermal cells.

More precisely, it is often the mold of the cuticle which, itself, carries this epidermal information. When the imprints display partial preservation of the original compression, we speak of "imprint/compression" or "adpression" (Martín-Closas and Gomez 2004).

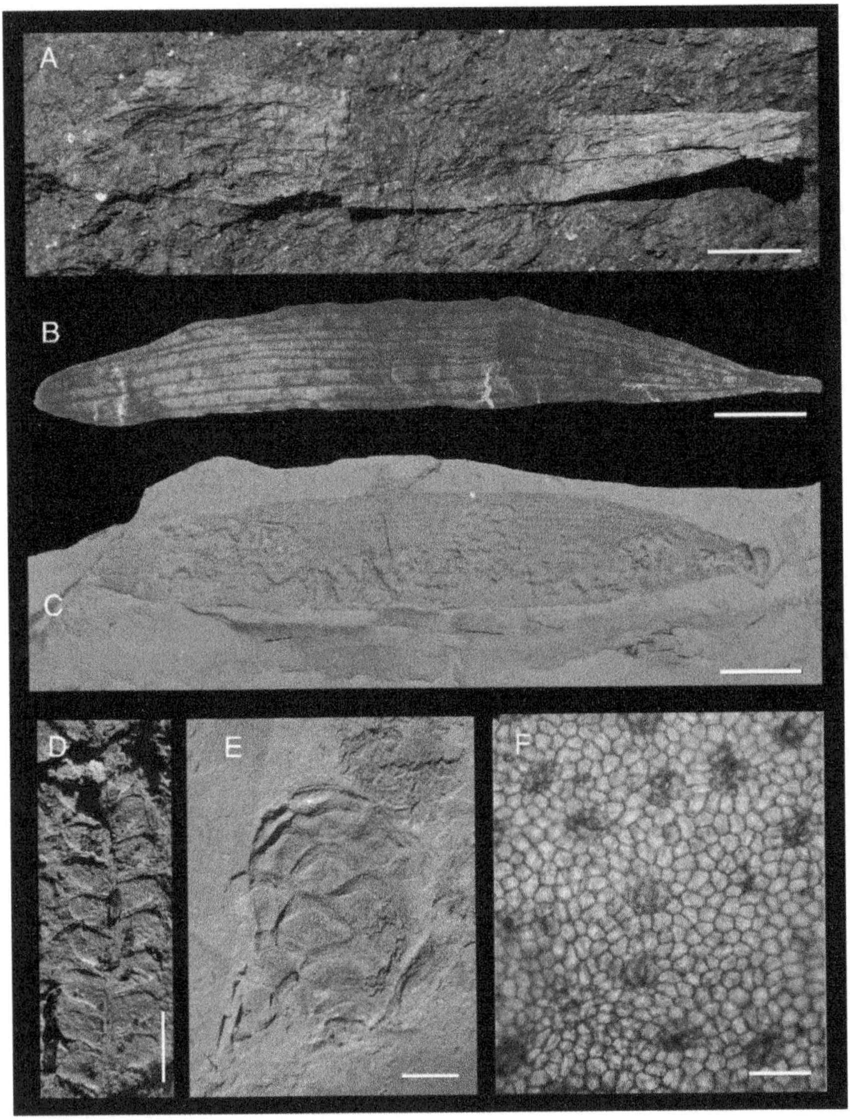

Plate 2.12. *Fossilized plants in laminated clay – clay facies dating from the Cenomanian, found in Charente-Maritime (western France)*

NOTE ON PLATE 2.12.– *A: cuticle in situ in lignite clay; B: cuticle removed from the clay, with the transparent view highlighting the fine preservation of the foliar nervation; C: leaf imprint in the clay without the cuticle having been preserved but with a thin, brownish, powdery carbonaceous residue; D: print of a fragment of a fern frond from which the carbonaceous fragments of cuticle are detaching; E: imprint of a cone in the clay, with a thin, brownish, powdery carbonaceous residue; F: high-magnification shot of a cuticle of* Eucalyptolaurus *which has finely preserved the imprint of the epidermal cells (collections: A to D, F: Néraudeau-IGR; E: É. Dépré; photos: A to D: D. Néraudeau; B and C: É. Dépré; E and F: J.-D. Moreau). Scale bars: A to E = 1 cm; F = 100 μm.*

2.3. Carbonatizations and calcitic crystallizations

The transformation of organic remains into calcareous fossils can take place early on, either during the organism's lifespan or shortly after its death; or else it can take place a very long way in the future: thousands or even millions of years after the organism's biological functions cease.

2.3.1. *Carbonatization by encrusting and impregnation*

When water circulates in an underground network in a calcareous landscape, it absorbs calcium carbonate in conditions of CO_2 pressure and pH that are considerably different from those on the surface. This calcium content is higher than that of surface water, and we speak of subterranean water in calcareous landscapes as calcareous water or hard water, characterized by an alkaline pH, though it is more acidic in water in a sandy or granite context. In addition, the solubility of the carbonates depends on the carbon gas content of the air; in a normal atmosphere at the surface, in a temperate region, water contains only 60 mg of lime per liter, it may contain four or five times more under the ground. Consequently, when subterranean water which has flowed through calcareous rocks reaches "open air" at the surface, it suddenly loses a large portion of its capacity to dissolve carbonates, and the excess precipitates out of solution.

Example of tufas and travertines

The consequence of the precipitation of calcium carbonate where subterranean water resurfaces from a calcareous underlayer is the rapid deposition of a calcareous film on all organic and mineral elements on the site of its surfacing, and downstream where that water flows while spreading carbonates. The organic elements which are subjected to calcareous encrustation may vary widely, from plants (roots, wood and stems, leaves and flowers, cones and fruits, moss, etc.) to animals (snail shells, arthropods, the bodies of small vertebrates, bird feathers, etc.) (Wilbert 1962). Over an extended period of time, the surface encrustation may become a deeper impregnation (see Box 2.3) of the empty spaces within the animal and plant microstructures, and lead to fine mineralization of the organic tissues. Finally, the carbonatized organism may become an essentially calcareous structure, having an organic framework which is either very limited or has been completely dissolved. The most remarkable example is that of fossils of tufas and travertines linked to petrifying sources. Fossilized tufas have been found at various sites in France, notably dating from the Paleocene/Eocene (Thanetian) in Sézanne, in Marne, north-eastern France, from the Oligocene in Charente (near Barbezieux), south-western France (see Plates 2.13A and 2.13B) and from the Holocene in Lozère (near Mende), south-eastern France (see Plates 2.13C to 2.13E) (De Saporta 1868; Néraudeau et al. 2013; Bourel et al. 2016). The plant remains that are most commonly preserved are mosses, followed by the leaves of angiosperms (see Plates 2.13A, 2.13C and 2.13D) and, to a lesser extent, fern fronds (see Plate 2.13B). All of them came from forests of leafy trees, which grew in calcareous ground near to outcrops. We may also find "petrified" winged seeds (see Plate 2.13E), nuts, shells of pulmonate gastropods, crawfish, bird feathers and insect wings.

We tend to speak of tufa for the calcareous encrustation that occurs in a context of normal ambient temperature ("temperate") and of travertine for calcareous encrustation occurring due to hot springs (Limondin-Lozouet et al. 2012).

Plate 2.13. *Fossils preserved in calcareous tufas*

NOTE ON PLATE 2.13.– *A and B: angiosperm leaf (A) and fragment of a fern frond (B) dating from the Oligocene, found in Touvérac, Charente, south-western France; C to E: angiosperm leaves (C and*

D) and samara (helicopter seed) (E) dating from the Holocene, found in Mende, Lozère, south-eastern France (collections: A and B: J. Drochon; C to E: APHPL; photos: A and B: D. Néraudeau; C to E: J.-D. Moreau). Scale bars: A, B, D and E = 1 cm; C = 0.25 cm.

Encrustation of the remains of organisms involves the deposit of a mineral (for example calcium carbonate) brought by a fluid which envelops these remains of animals or plants. The context in which this takes place is often described as a "petrifying spring". The process begins with the deposition of a mineral film over the remains of the organisms, and after the dissolution of the organic remains, there remains an imprint of those remains (in inverse relief). If exposure to mineralization continues over time, the mineral *impregnation* penetrates the porosity of the tissues or the skeleton, hardens them, and may form a copy equal in volume to the original remains.

Box 2.3. *The concepts of encrustation and impregnation*

The formation of tufas is closely linked to the composition of bacterial biofilms which develop on the surface. These fine films formed of cyanobacteria absorb CO_2 from the water by photosynthesis (like the chlorophyllic plants in direct contact with water). Thus, the biofilm helps to increase the biochemical precipitation of calcite on the edges of the tufa fields and travertines (Limondin-Lozouet et al. 2012).

2.3.2. Carbonatization by percolation and recrystallization

In the case of calcitization, the original state of the fossil organism is generally organic, but it may also contain mineralized parts. The different phases of mineralization can be observed and analyzed in thin layers (see Box 2.4). The most remarkable case is that of calcretes. Calcrete is a pedogenetic cementation of calcium carbonate due to a level of water infiltration loaded with limestone (Vogt 1984; Carcinelli et al. 1989; Giresse and Martzluff 2016). This cementation in a hydromorphous environment may be relatively planar on the scale of a layer of varying thickness, or very localized on the scale of small fluffy nodules (sometimes known as "limestone dollies"), whose nucleus may be a plant or animal element. Carbonate precipitation is generally linked to bacterial activity – notably by cyanobacteria (Freytet and Plaziat 1965).

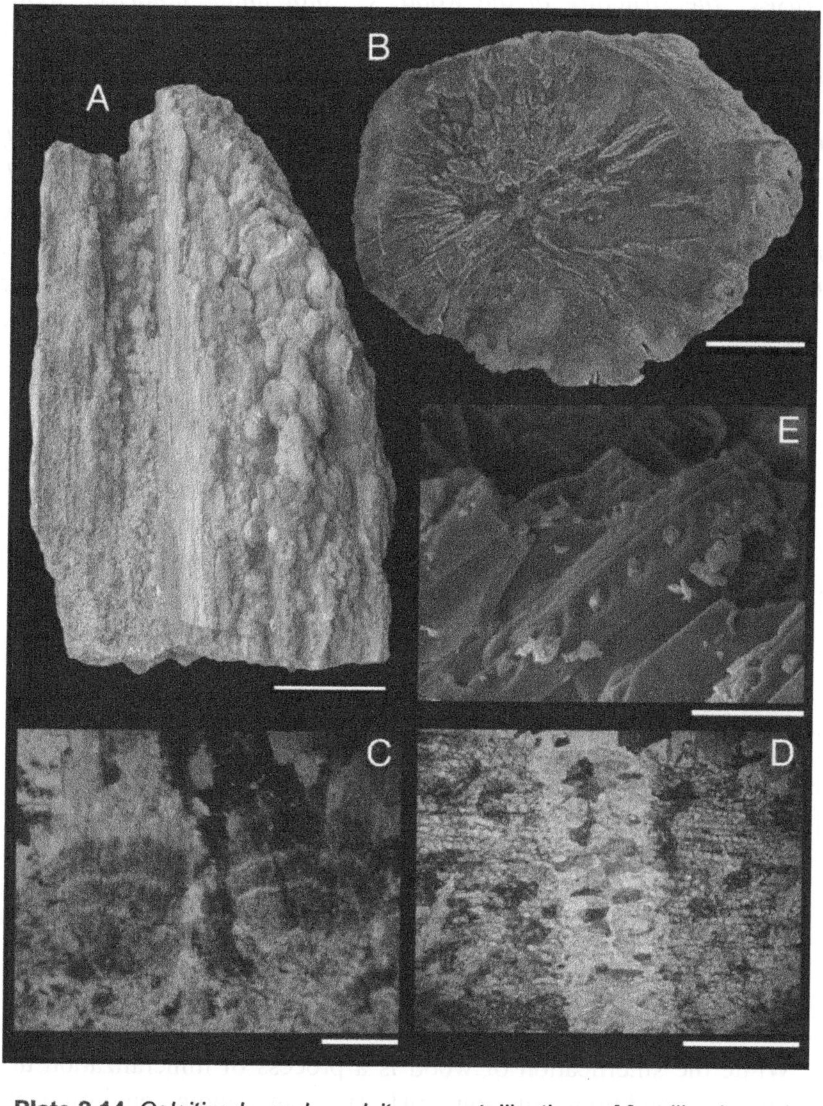

Plate 2.14. *Calcitized wood – calcite recrystallizations of fossilized wood, dating from the Berriasian, found in Cherves-Richemont, Charente, south-western France*

NOTE ON PLATE 2.14.– *A: fragment of a calcitized branch bearing a nodular encrustation of spherules of calcite, of microbial origin; B: polished section of calcitized wood showing that the calcite largely "erases" the structure of the wood; C: shot under polarized light, showing a thin sliver of calcitized wood, where the original cell networks, shown in brown, are rare, with the microstructure of the wood being largely replaced by shafts of microcrystals of calcite; D: shot under polarized light, showing a thin sliver of calcitized wood, where the original cell networks, shown in brown, are well preserved, with calcite microcrystals (polarizing in grayscale) replacing them in groups or in a crack (collections: A and B: J. Drochon; C and D: Néraudeau-IGR; E: Université Claude Bernard Lyon; photos: A to D: D. Néraudeau; E: M. Philippe). Scale bars: A = 3 cm; B = 5 cm; C and D = 0.25 mm; E = 20 μm.*

The thin sections are very thin slivers of rock (on average of the order of 20 to 30 μm in thickness), glued to strips of glass, which are designed for observation under a binocular lens or an optical microscope. These rocky plates need to be sufficiently fine to be translucent and allow light to pass through, and thus observed under transparency under high magnification. They are produced by gradually thinning a fragment of rock (by erosion on a turntable covered with abrasive material), previously glued to a glass slide, until it is reduced to a thin section. This technique enables us to observe not only the microstructure of the rock, but also to discover and identify microfossils contained within it (for example foraminifera), or the histological details of a macrofossil (for example plant matter, bones). It should be noted that observing these thin sections under polarized light reveals the nature of the minerals making up the fossilized microremains and their sedimentary surroundings (see Plate 2.14C and 2.14D).

Box 2.4. *Technical preparation of thin sections*

Example of calcitization of wood

While the silicification of wood is a process of mineralization that occurs in conditions of low pH (i.e. an acidic environment), calcitization, for its part, is rather typical of a context with a high pH (an alkaline environment). In addition, wood calcitization more generally occurs at moderately warm temperatures, rather than in extremely hot temperatures. Finally, it often involves microbial activity which mineralizes the organic tissues in a centripetal fashion, with an encrustation of the surface of the wood by banded carbonate spherules with a bacterial "seed" (see Plate 2.14A), extending into the ligneous tissues by a polymorphous recrystallization of the vessels in the wood (see Box 2.5): monoblastic, oligoblastic or pluriblastic (see Plate 2.14C and 2.14D) (Néraudeau et al. 2022). In spite of the modification of the macroscopic appearance of the wood (see Plate 2.14B), the cellular network may be more or less well preserved locally, with the cell walls retaining their organic nature and a brownish color which, under polarized light, contrasts with the calcite crystals which have substituted the organic materials elsewhere. Under a scanning electron microscope, the best preserved wood vessels enable us to identify the type of tree (see Plate 2.14E). Wood calcitization can be observed both in a terrestrial environment and in the coastal marine environment. On the Berriasian (Lower Cretaceous) site at Cherves-Richemont, Charente (south-western France), the wood was carbonatized in a swampy or lacustrine context.

> The crystallization of a mineral within – or in place of – a cellular network can occur in three ways: *monoblastic*, when a crystal corresponds to a single cell; *oligoblastic*, when a crystal encapsulates a small number of cells; or *pluriblastic*, when the same crystal affects a large number of cells.

Box 2.5. *The concepts of monoblastic, oligoblastic and pluriblastic recrystallizations*

2.3.3. *Geodes and calcitic outgrowths*

In addition to the microcrystalline recrystallizations mentioned previously, calcite has the peculiar quality of forming macrocrystals in the natural cavities of various fossilized organisms. The carbonated fluids which percolate into these cavities, in contexts of marl/calcareous rocks, typically crystallize to form geodes (see Figure 1.3B'). This is often observed in fossils of ammonites or nautiloids, where calcite crystals coat the walls of the internal chambers and form geodes between two chambers when a void remains (see Plates 2.15A and 2.15C), or fill the whole of the space by centripetal crystallization (see Plate 2.15B). Comparable cases can also be observed in the internal cavities of sea urchins, or in the voids separating the valves in bivalves, whether they are "classic" lamellibranchs (see Plate 2.15E) or rudists (see Plate 2.15D).

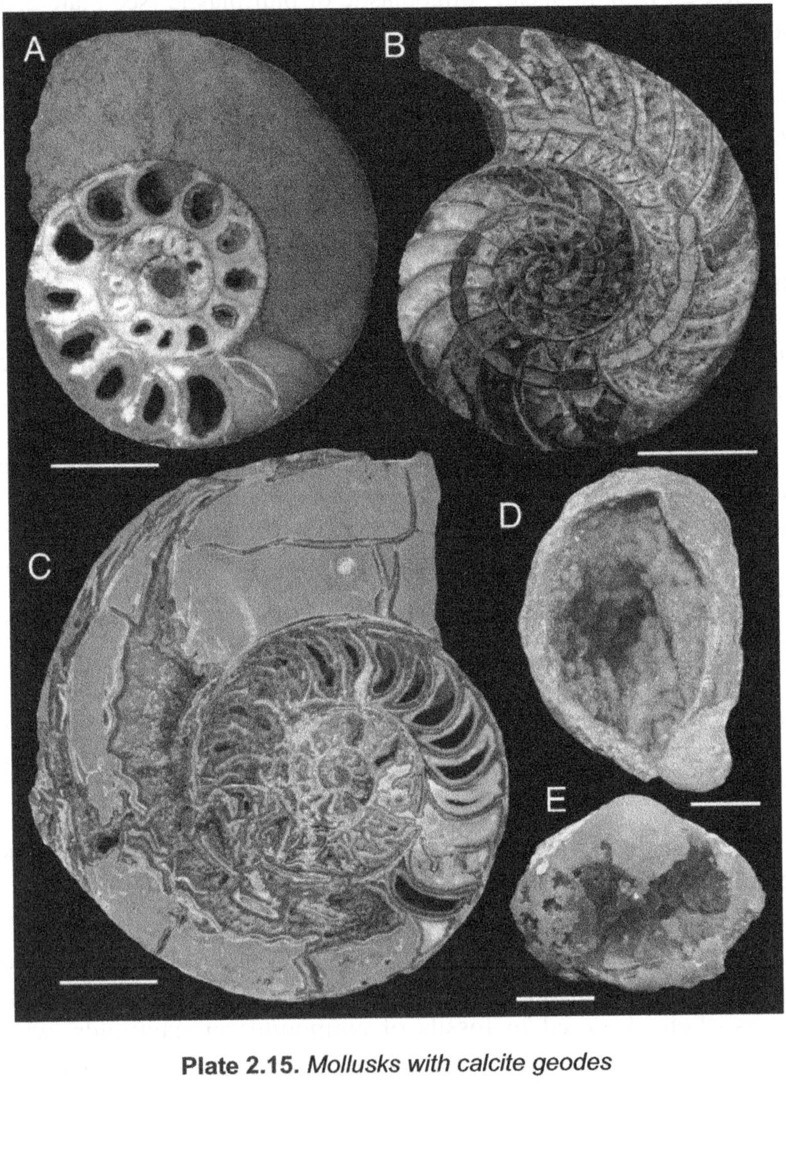

Plate 2.15. *Mollusks with calcite geodes*

NOTE ON PLATE 2.15.– *Polished sections of ammonites (A and C) and nautiloids (B) showing different recrystallizations of calcite, constituting small geodes on the surface of the shell's internal walls (A and B) or completely filling certain cavities (B and C). A: ammonite dating from the Oxfordian, found in Madagascar; B: nautiloid dating from the Oxfordian, found in Madagascar; C: ammonite* Asteroceras stellare *dating from the Sinemurian, found in Dorset, Britain; D: hollow valve of a rudist of the type* Polyconites, *dating from the Cenomanian, found on Île Madame (Charente-Maritime, south-western France), carpeted with small crystals of calcite; E: bivalve dating from the Mio-Pliocene, found in Iceland (collections: A and B: P. Miramand; C: IGR; D: Néraudeau-IGR; E: É. Dépré; photos: A and B: T. Bouyer; C to E: D. Néraudeau). Scale bars: A = 5 cm; B = 3 cm; C = 10 cm; D = 5 cm; E = 2 cm.*

Plate 2.16. *Sea urchins with geodes or crystalline extensions of calcite*

NOTE ON PLATE 2.16.– *A: irregular cassiduloid sea urchin* Catopygus columbarius, *dating from the Cenomanian, found in Tonnay-Charente, Charente-Maritime, south-western France, with extensions of calcite crystals developing from each plate of the test, growing outwards; B: regular sea urchin, dating from the Aalenian, found in Poitou, central France, with extensions of calcite crystals developing from each plate of the test, growing outwards; C: irregular cassiduloid sea urchin* Catopygus columbarius, *dating from the Cenomanian, found in Piédemont, Charente-Maritime, south-western France, with extensions of calcite crystals developing from each plate of the test, growing inwards; D: irregular spatangoida sea urchin* Periaster undulatus, *dating from the Cenomanian, found in Île d'Aix, Charente-Maritime, south-western France, with extensions of calcite crystals developing from each plate of the test, growing inwards (collections: A, C and D: Néraudeau-IGR; B: A. Beauvois; photos: D. Néraudeau). Scale bars: A and B = 0.5 cm; C and D = 1 cm.*

More rarely, crystals of neoformed calcite "extend" the external or internal face of the test of fossilized sea urchins (see Figure 1.3C'). Generally, the various plates making up a sea urchin are preserved in the form of calcite monocrystals, as is every radiole (spine), but the fossil is identical in appearance to the unfossilized specimen.

However, in "aerated" sedimentary contexts (shelly sand, calcareous sand), the presence of a carbonated fluid in the porosity of the sediment "nourished" calcitic growth, enabling each crystal/plate of the echinoderm to grow beyond its normal size, with "houndstooth" calcite crystals constituting outgrowths which stand out from the test's external surface (see Plate 2.16A and 2.16B) or internal surface (see Plate 2.16C and 2.16D).

Finally, even if they are not crystals of calcium carbonate but of calcium fluoride (CaF_2), cubic crystallizations of fluorite have been observed on fossilized shells in the UK (Dirtlow Rake, Derbyshire) or filling the internal cavity of bivalves in Germany (Caaschwitz, Thuringia).

2.3.4. *Crystalline fossils formed of carbonates*

Aside from the previous cases of mass carbonatization, there are examples of carbonated replicas, when calcite crystals have filled the void left by a dissolved fossil during a late mineralization event. This case is often encountered with calcite fossils, such as the gastropod and bivalve mollusks dating from the Lower Jurassic (Hettangian, –201 to –199 Ma) from the coast in Vendée (Jard-sur-Mer, Talmont-Saint-Hilaire, western France), calcitized during the Cretaceous by the percolation of a fluid charged with calcium carbonate (see Plate 2.17A and 2.17B) (Bouat 2023).

In the case of calcite fossils, they are replicas formed of limestone (non-crystalline) in a first diagenetic episode, then dissolved during a second episode, with the void thus formed finally being filled by the addition of calcite, closely mirroring the volume and ornamentation of the original fossil.

Note that, in the same carbonated context, and following the same process of diagenetic replacement, large crystals of halite (i.e. kitchen salt, sodium chloride, NaCl) may retain a perfectly cubic form when they become calcite monocrystals. These pseudomorphs of halite may be described as fossil salt crystals (see Plate 2.17C and 2.17D), in the same exact position where they were formed by evaporation, and thus become excellent markers of local paleosalinity. There are exceptional cases of fossilizations of stromatolites, including cubic crystals of calcitized salt, which demonstrate the high level of salinity of the context in which these microbial bioconstructions developed (see Plate 2.17D).

Rare cases of pseudomorphs in pink manganese carbonate, otherwise known as rhodochrosite ($MnCO_3$), are also known – particularly certain mollusks from the Pliocene, found in the Kerch Peninsula, Crimea (Ukraine) (see Plate 2.17E and 2.17F) (Golubovskaya 2001; Chukanov 2005). These replicas, geodes or encrustations by rhodochrosite, result from the hydrothermal epigeny of pre-existing fossils with a manganese-rich fluid.

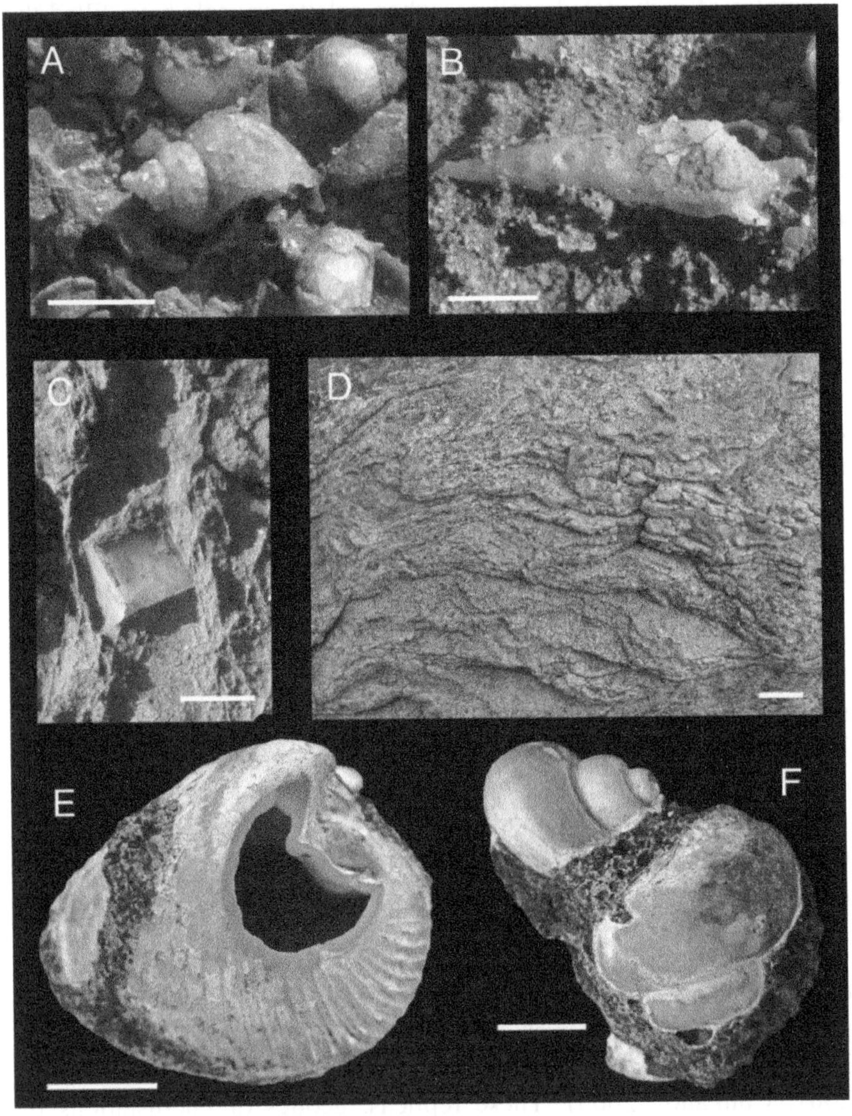

Plate 2.17. *Fossils with epigeny in calcite or rhodochrosite*

NOTE ON PLATE 2.17.– *A and D: Cretaceous epigeny of calcite deposits from the Lower Jurassic (Hettangian) from the coast of Talmont-Saint-Hilaire, Vendée, western France. A and B: gastropods totally epigenized into calcite, with a crystalline appearance; C and D: cubic crystals, originally formed of halite salt, epigenized into calcite within the sediment (C) or in stromatolitic domes (D); E and F: rhodochrosite fossils from the Kerch Peninsula, Crimea (Ukraine): E: bivalve* Arcicardium; *F: gastropods (collections: A to D: in situ; E and F: B. Hogrel; photos: D. Néraudeau). Scale bars: A and B = 0.25 cm; C to F = 1 cm.*

2.4. Pyritization, limonitization and siderite

Various types of fossilization use iron as a support element for the mineralization, be it by means of sulfides or oxides. The most frequent process is pyritization, with iron sulfides, but there are also cases of fossilization with hematite, therefore with oxides.

2.4.1. *Pyritization*

Pyrite is an iron disulfide (FeS_2), which forms when iron, contained in the sediment or in the tissues of the organisms, comes into the presence of hydrogen sulfide (H_2S), generally formed by the decomposition of organic matter by certain sulfur bacteria.

Indeed, certain bacteria produce hydrogen sulfide by breaking down proteins, with sulfur-reducing species decomposing sulfates into H_2S. In parallel, other bacteria reduce ferrous hydroxides from the organic matter and release ferrous ions, Fe^{3+}.

The combination of hydrogen sulfide with ferrous ions from the same anoxic context subject to bacterial activity leads to the precipitation of pyrite (Borkow and Babcock 2003).

Pyrite precipitates across a wide range of pH values and redox potential Eh (see Box 2.6.), as long as a sulfate compound is present.

However, pyrite formation is favored by an environment with a redox potential less than +0.25 (see Figure 2.4) (Mustoe 2018).

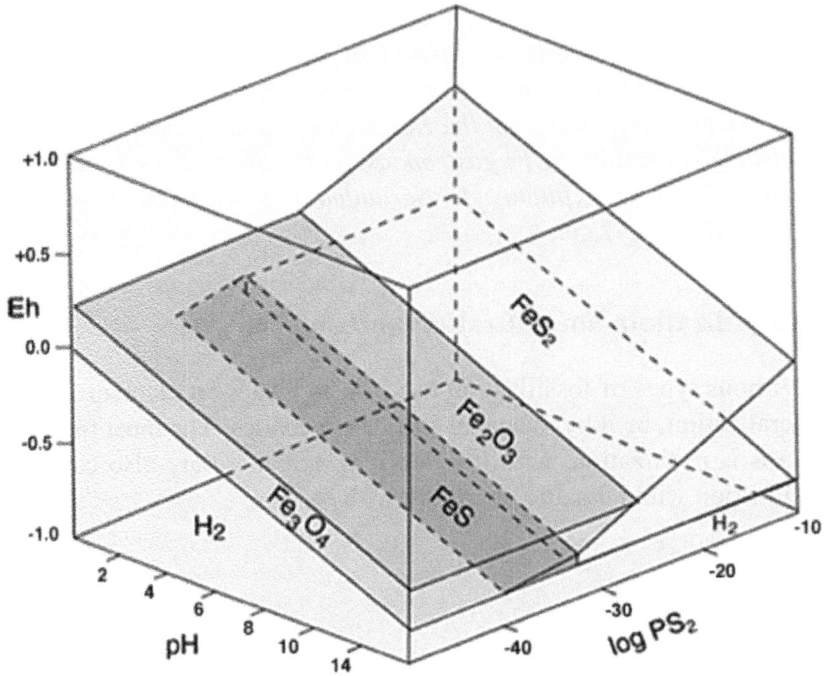

Figure 2.4. *Pyrite formation*

NOTE ON FIGURE 2.4.– *Phase diagram for sedimentary ferrous minerals. In a reductive environment, pyrite precipitates at widely varying values of pH and redox potential Eh, as long as sulfates (log PS_2) are present in relative abundance. Other iron-rich sulfides, such as pyrrhotite (FeS), precipitate when the sulfur content is very low. In an oxidative environment, we tend to see the precipitation of iron oxides such as magnetite (Fe_3O_4) or hematite (Fe_2O_3) (taken from Mustoe (2018)).*

The *redox potential* (Eh) is a measure of the tendency of an environment to oxidize or reduce a substrate. An aerobic substrate, which is an oxidizing environment, has an Eh of +800 mV, whilst an anaerobic substrate, which is a reducing environment, has a negative Eh, which may be as low as –300 mV.

Box 2.6. *The concept of redox potential (Eh)*

In present-day environments, coastal mudflats (bay mud, pitch sand), marshy or freshwater (peat) often constitute fine, anoxic sediments, ranging from olive green at the surface to blue-black at greater depth, and releasing a characteristic sulfurous smell of rotten eggs, due to the H_2S. The organisms buried in these sediments are soon deprived of oxygen and are thus protected from certain decomposers, whilst being in the presence of a large quantity of particulate organic matter, sulfides and reducing bacteria (Billon 2001). Their preservation may therefore be limited to carbonaceous "mummification", as often happens for plants which accumulate in these environments. However, as a result of bacterial action, iron sulfides may fill the porosity of the tissues or shells by permineralization, or fill the internal cavities of the organisms and even come to substitute the majority of the anatomical architecture.

It is therefore possible to see either the addition of pyrite to organic remains, bones or shells, or their substitution by iron sulfides, or indeed both at once or successively. The most commonplace pyrite substitutions are those of carbonates, or of the organic material, by exogenous or endogenous pyrite. Exogenous pyritization is the most common process, with iron sulfides coming from the surrounding medium (sediment or inflowing fluids) around the organism being fossilized, and there are cases of endogenous pyritization, in which the iron and sulfur may come from the mineral parts, tissues or fluids of the organism itself.

The substitution of the original mineral (carbonate, phosphate, etc.) from a shell or bone by pyrite may produce a total or partial replica of these skeletal parts (see Plates 2.18, 2.19A and 2.19B). Thus, the percolation of an iron-sulfide-rich fluid may lead to the partial replacement of a carbonate shell, with pyrite being found alongside the remains of the shell that have not been epigenized (see Plates 2.19A and 2.19B).

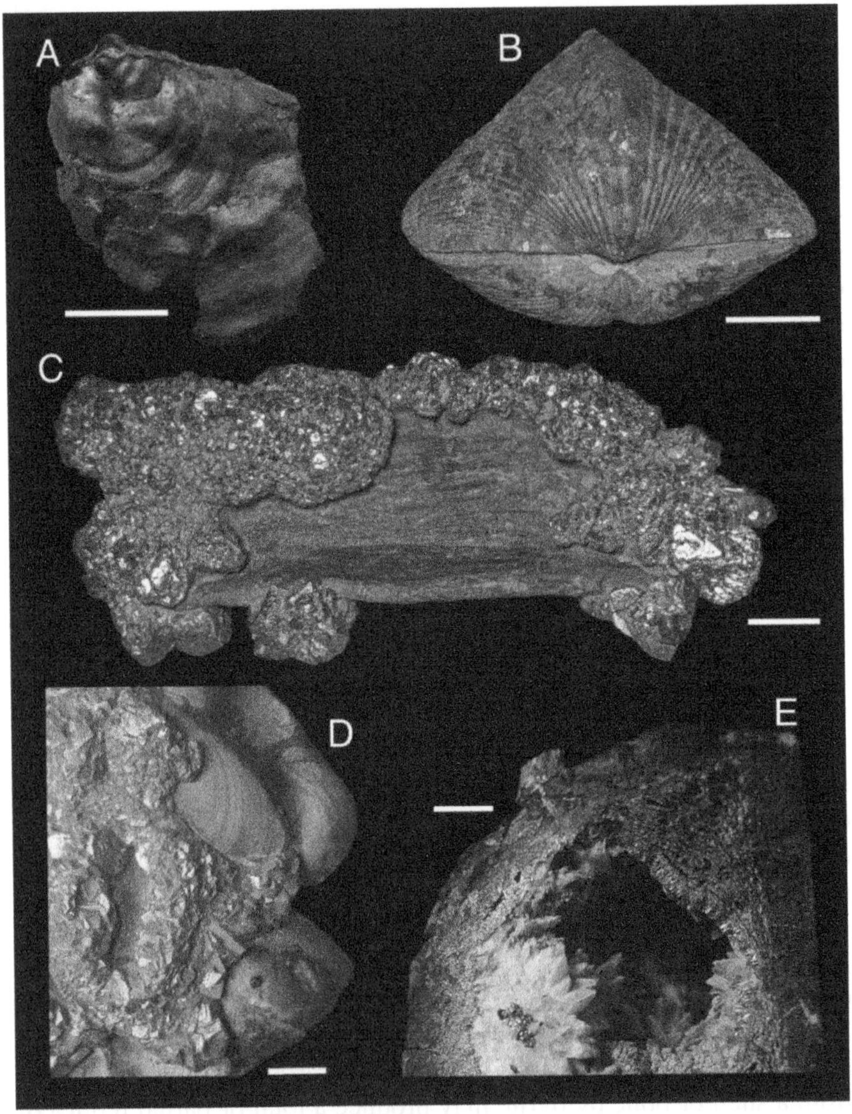

Plate 2.18. *Various fossils of pyrite or marcasite*

NOTE ON PLATE 2.18.– *A: oyster of the genus* Curvostrea, *dating from the Cenomanian, found in Fouras, Charente-Maritime, south-western France; B: brachiopod of the type rhynchonella, origin unknown; C: fossilized wood of lignite, whose porosity is pyritized, with outgrowths of marcasite crystals on the surface, origin unknown; D: pyritized bivalves, Missouri, USA; E: cross-section of a bone from a theropod dinosaur, dating from the Berriasian, found in Angeac, Charente, south-western France, with a porous medullary part recrystallized with pyrite, and the wall of the central cavity covered with a centripetal pyrite layer, itself serving as a support for a later crystallization of calcite in "houndstooth" form (collections: A, E: Néraudeau-IGR; B to D: J.-C. Le Sager; photos: A and E: D. Néraudeau; B to D: M. Guillo). Scale bars: A and C = 1 cm; B, D and E = 0.25 cm.*

The addition of exogenous pyrite, when it fills the whole of the volume of the original organism, often leads to the formation of a pyritous replica which replaces both the shell and its internal cavities (see Plates 2.18A, 2.18B and 2.18D). However, filling with pyrite may be limited to the internal cavities, in which case, after the later dissolution of the shell, the final fossil is limited to a pyritous internal cast of the original organism (see Plate 2.19C to 2.19E).

In a context of pyritization, fossils may also show crystalline outgrowths, either on the shells or bones or, still more commonly, on the ligniteous wood, largely carbonaceous, but pyritized to a greater or lesser degree in its microstructure (see Plates 2.18C to 2.18E and Plate 2.19A). In most cases, the outgrowths of iron sulfide from the fossils correspond to marcasite. This mineral has the same chemical composition as pyrite, but unlike pyrite (which forms cubic crystals), it crystallizes in orthorhombic form. Marcasite encrustations are highly sensitive to oxygen and moisture in the air, and disintegrate fairly quickly when they are exposed, whereas true pyrite crystals are more stable.

We do not speak of exceptional preservation when the pyrite merely forms an internal cast of a shell, bone or plant stem. However, the mineralization of fossils by pyrite, or pyritization, is a process which can sometimes preserve fragile organic elements, such as phanera or soft tissues. The chemical mechanisms governing this type of remarkable fossilization are always favored by lack of oxygen (for example in amber, see Plate 2.20), the presence of organic material and bacteria, in a context including varying levels of sulfur and iron.

Plate 2.19. *Pyritized ammonites*

NOTE ON PLATE 2.19.– *A: ammonite fossil, from the Albian, Madagascar, comprising an internal cast made from brown calcite (Ca), largely covered by the preserved shell (Co), but locally encrusted with pyrite (Py); B: the shell (Co) of an ammonite, compressed and partially recrystallized by pyrite with a metallic sheen (Py), from the Callovian, Chippenham, UK; C and D: pyritous ammonites dating from the Toarcian, found in Barjac, Lozère, south-eastern France, having lost a little (C) or a larger amount (D) of their metallic sheen; E: pyritous ammonite of the genus* Harpoceras *in the process of oxidation, with pyrite ceding its place at the surface to a reddish film of hematite, from the Toarcian, Le Clapier, Aveyron, south-eastern France (collections: A: É. Dépré; C and D: J.-D. Moreau; B and E: IGR; photos: A, B and E: D. Néraudeau; C and D: J.-D. Moreau). Scale bars: A and B = 5 cm; C to E = 1 cm.*

Example of ammonites with a pyritized internal cast

The best-known pyrite fossils are probably the pyritous ammonites from the Lower Jurassic, present in numerous deposits all over the world, such as in Aveyron and Lozère, in south-eastern France (Monestier 1931, 1934; Guex 1972; Jattiot et al. 2015a). They are generally internal casts made of pyrite, for ammonites whose shell was dissolved during diagenesis, and only the internal cavities, filled with iron sulfides, constitute the fossil. The chambers, with very well defined lines ("veined"), are then clearly visible (see Plates 2.8C to 2.8E). These total pyritizations of ammonites occur in clay/marl contexts which are decarbonated, anoxic and rich in the circulation of iron sulfides. However, it sometimes happens that the ammonites are fossilized and keep their shell, and that shell is either completely or partially pyritized (see Plate 2.19A and 2.19B). There may be cases of complex mineralization, combining an internal cast made of calcite, widespread preservation of the carbonate shell, but also some encrustations or replacements at certain points by pyrite (see Plate 2.19A). Indeed, the shell may remain present in parts in its original carbonate form, but be replaced by pyrite elsewhere. These partial pyritizations can only happen if the surrounding sediment is still rich in carbonates, and if the percolation of fluids containing iron sulfides is fairly limited, in a slightly anoxic context.

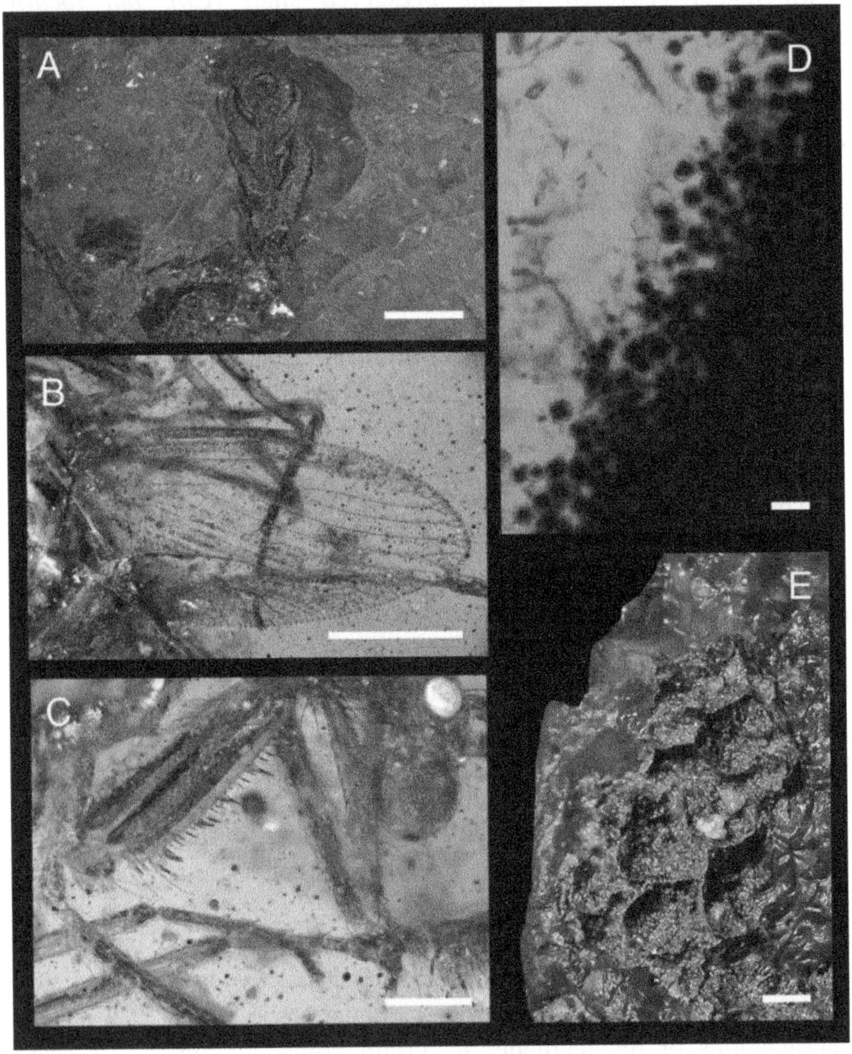

Plate 2.20. *Pyritized organisms as inclusions or encrustations in amber*

NOTE ON PLATE 2.20.– *A: the end of a conifer branch, from the Cenomanian of Fouras, Charente-Maritime, south-western France; B and C: neuroptera Rhachiberothidae fossilized in amber from the Cenomanian, Noije Bum Hill, Kachin State, Myanmar; B: wing with microcrystals of pyrite developed on the alar nervations; C: compound eye and feet with internal filling by pyrite; D: microcrystals of pyrite developed in a coprolite (right) and cyanobacterial filaments (left) in Cenomanian amber, Archingeay-Les Nouillers, Charente-Maritime, south-western France; E: pyritous oysters encrusting a nodule of amber, from the Cenomanian, Fouras, Charente-Maritime, south-western France (collections: A, D and E: IGR; B and C: MNHN; photos: A: V. Perrichot; B and C: C. Jouault; D: V. Girard; E: D. Néraudeau). Scale bars: A and E = 5 mm; B = 1 mm; C = 0.24 mm; D = 10 µm.*

Example of pyritized dinosaur bones

The pyritization of vertebrate remains may occur by combination of the iron produced by the degradation of hemoglobin and sulfur from the hydrolysis of proteins. This pyrite, which is known as "endogenous", is deposited in the interstitial spaces in the bones and the voids left by the decomposition of the cells. A famous example of bone pyritization is that of the Bernissart *Iguanodon* (Norman 1980; Godefroit and Leduc 2008) – a herd of dinosaurs unearthed in Belgium in 1878. A comparable example was observed in the bones of theropod dinosaurs from the Berriasian deposit in Angeac-Charente, south-western France, pyritized in the medullary cavity which contained the marrow. In this deposit, crystallizations of exogenous pyrite, drawn from the surrounding marl/lignite context, are added to the endogenous pyrite, in the tubular part of the bones. Finally, geodes of calcite, from the carbonated fluid from the overlying calcareous rock (calcareous shingle dating from the Pleistocene), are sometimes also added to the pyrite encrustation (see Plate 2.18E).

Example of pyritization in amberiferous lignites

Amberiferous lignites are known for cases of amber which accumulates and constitutes a useful material for exceptional fossilization (see section 2.6). However, in amberiferous lignites, the pyritization may involve additional and varied processes of remarkable preservation. Thus, in amberiferous lignites (for example the sites dating from the mid-Cretaceous in Charente-Maritime, south-western France, or from the Eocene in Oise, Paris region), we find, side by side in the same facies:

1) conifer wood (*Agathoxylon*, *Brachyoxylon*, etc.), whose porosity is filled with pyrite or marcasite, which develop as crystalline outgrowths on the surface of fragments of trunks or branches (see Plate 2.18C);

2) completely pyritized oysters, having the appearance of "metal shells", attached to the wood which floated to the coast, before being deposited and fossilized there (see Plate 2.18A);

3) pyritized oysters encrusting nodules of ambers (see Plate 2.20E);

4) arthropods or plant remains in amber that have been transformed into pyrite to a greater or lesser extent, these too having a silvery metallic sheen (see Plates 2.20A to 2.20C).

In this latter case, the iron sulfides penetrated into the core of the amber through microfractures, themselves pyritized in the form of "pyrite microlayers". The iron sulfides reached certain air bubbles trapped in the resin and formed small pyritous silver globules. Ferrosulfide fluids, drawn from the ligniteous deposits, infiltrate the amber through microfractures or the natural tubular structures of animal inclusions (antenna, feet, alar nervations), plant inclusions (vessels in wood, foliar nervations) or mycelial inclusions (filaments) (see Figure 2.5). It sometimes reaches micro-organisms trapped in the resin at the same time as the arthropods, such as single-celled algae (diatoms), or protists (testate amoebae). Finally, the anoxic context of the ligniteous clays favored the development of cyanobacteria and microscopic fungi, whose mycelial filaments proliferated in the resin in the process of fossilization, forming a crust made up of their filamentary network at the edge of the amber nodules. These filaments themselves may sometimes be mineralized by cubic microcrystals of pyrite, within the amber (see Plate 2.20D).

a

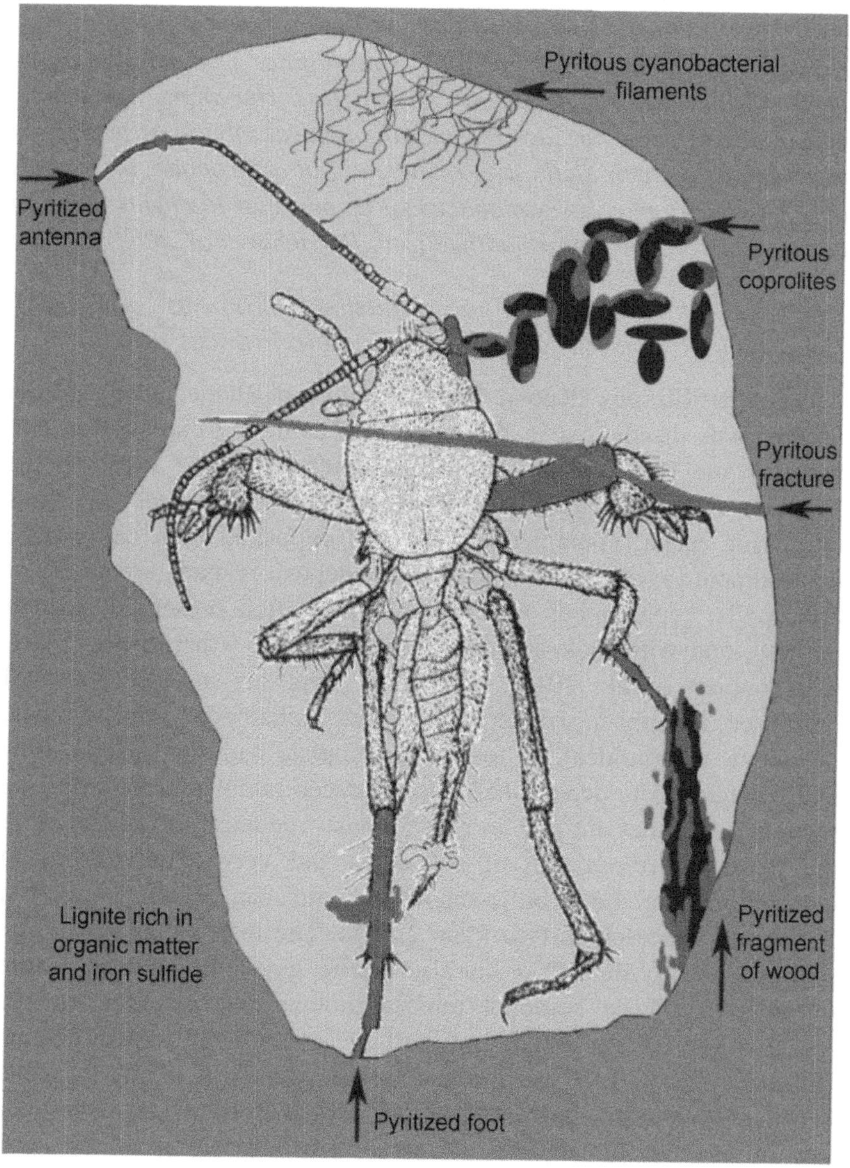

Figure 2.5. *Formation of pyrite in amber*

NOTE ON FIGURE 2.5.– *The nodules of amber (orange) are generally embedded in lignite-containing clays or marls (shown in blue), very rich in organic material, but also iron sulfide. The percolation of these sulfides into the sediment can penetrate the amber (green) by means of fractures, but also through the natural tubular structures of the organisms, such as the feet or antenna of arthropods, the vessels of the wood fragments, cyanobacterial or mycelial filaments, or even beads of microcoprolites (modified from Perrichot et al. 2002).*

Example of the Callovian deposits (–165 to –161 Ma), Voulte-sur-Rhône (Ardèche, south-eastern France)

The fossiliferous deposit in La Voulte-sur-Rhône, south-eastern France, was discovered in the context of a mining exploitation that began in the 18th Century. The digs, carried out in the 1980s by the Muséum d'Histoire naturelle de Paris, yielded exceptional specimens of marine fauna, contained in the marl deposited in a calm, deep environment (over 200 m in depth). The deposit is made up mainly of arthropods, cephalopods and echinoderms, often preserved in their entirety, and with their anatomical connections (Charbonnier 2009; Charbonnier et al. 2014). The soft parts are remarkably well preserved, notably in pyritized form. The cephalopods (belemnoidea, coleoidea, theuthoidea) are fossilized complete with their heads, arms and tentacles. The deposit has also produced nodules of siderite (see section 2.4.3) that are rich in crustaceans – notably, Coleiidae. With the frequent preservation of soft parts and very good anatomical connection, the La Voulte-sur-Rhône site can be considered a Konservat-Lagerstätten (see Plate 2.21A). The fossils contain a large proportion of pyrite (50%), but also calcite, gypsum, and even apatite or galenite. These complex mineralizations may account for the preservation of the details of certain anatomical microstructures (Fischer 2003). The exceptional fossilization was made possible because the marine sludge and its interface with the water were very low in oxygen. In addition, the context of the burial of the dead organisms was highly rich in minerals in solution, lending itself well to the early mineralization of the organic structures in connection with bacterial activity.

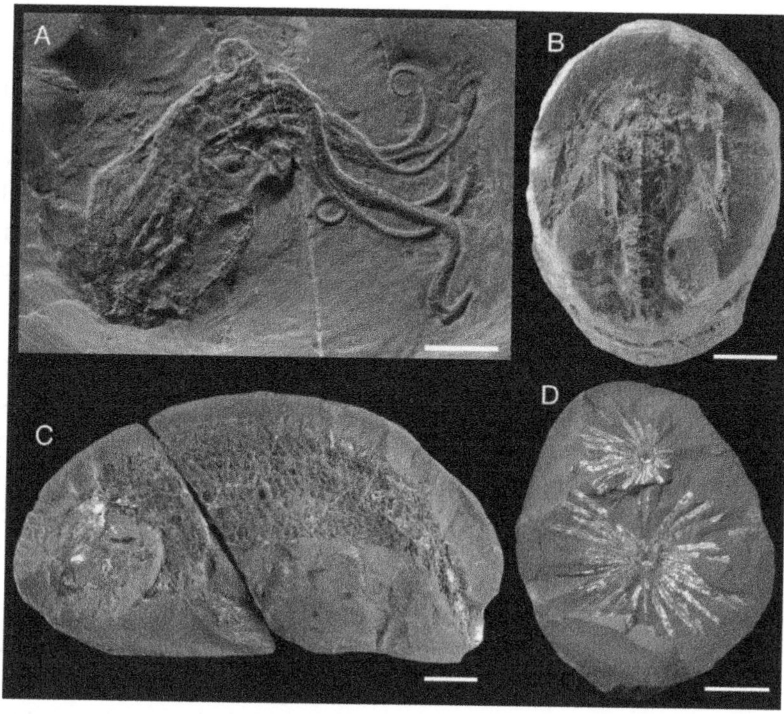

Plate 2.21. *Siderite fossils*

NOTE ON PLATE 2.21.– *A and B: pyritized fossils in fine marl sediments (A) or in siderite nodules (B), found in La Voulte, Ardèche, south-eastern France. A: coleoid cephalopod* Proteroctopus ribeti, *with pyritization of the soft tissues; B: decapod crustacean Polychelida preserved in a siderite nodule; C and D: fossils preserved in siderite nodules, dating from the Carboniferous, found in Montceau-les-Mines, Saône-et-Loire, central France: (C) actinopterygian fish; (D)* Annularia stellata *(collection: MNHN; photos: A: P. Massicard; B: P. Loubry; C and D: N. Podvigne). Scale bars: A = 2 cm; B = 0.25 cm; C and D = 1 cm.*

2.4.2. *Fossils made of iron oxides or hydroxides*

In iron-rich sedimentary environments and subject to significant hydration, the ferrugination of certain fossils may take place in the form of iron oxides (Kosakevitch 1983). Hematite, a compound of iron oxide, has the formula Fe_2O_3 (see Figure 2.4). It is a fairly common mineral, in a metamorphic context and in a sedimentary context, generally dark in color – gray, black or brown, with red highlights.

In hydrated form, iron oxides often become limonite, $FeO.H_2O$, corresponding to a mass of microcrystalline iron hydroxide, reddish brown (when "fresh") or ochre (when altered). Another hydrated iron oxide, goethite (FeO(OH)), has a shiny black color, generally without the red tinges of hematite. Among all these iron oxides or hydroxides, it is not always easy to clearly determine which is involved in certain cases of fossilization through epigeny.

The case of hematite is often pointed to for the fossilization of certain ammonites which – unlike pyrite, gold or silver ammonites with metallic reflections – are more matte and darker, with a reddish brown-black color (see Plate 2.22A and 2.22B). The internal casts of ammonites (i.e. with visible chambers) which were filled by an iron-rich matrix would be mineralized in oxides such as hematite in an oxidating context, and in pyritous sulfates in a reducing context.

Note that bivalve mollusks can also lead to replicas in pyrite (see Plate 2.22C) or hematite. In sedimentary contexts subject to water percolation or to temperature elevation, pyritous ammonites can potentially undergo epigeny to hematite (Maurel 1964; Pons 2015).

Aside from the frequent case of ammonites, more exceptional cases of hematite fossilization have been discovered, for fossils of varying geological ages, such as segments of axes of *Archaeosigillaria* dating from the Devonian (–419 to –359 Ma) in Morocco (see Plate 2.22E).

Oxidation of pyrite to hematite:

$$4\ FeS_2 + 15\ O_2 + 8\ H_2O \rightarrow 2\ Fe_2O_3 + 8\ H_2SO_4$$

Plate 2.22. *Various fossils composed of hematite (A to C, E), or limonite (D)*

NOTE ON PLATE 2.22.– *A: ammonite* Phylloceras heterophyllum *from the Lower Jurassic, La Verpillière, Isère, south-eastern France; B: ammonite of the genus* Amaltheus *from the Lower Jurassic, found in central France; C: bivalve* Leda rostralis *from the Toarcian, found in the Millau region, Aveyron, south-eastern France; D: limonitized wood from the Cenomanian, Fouras, Charente-Maritime, south-western France; E: axis of a plant from the Devonian, of the genus* Archaeosigillaria, *found in Morocco (collections: A to D: IGR; E: B. Hogrel; photos: D. Néraudeau). Scale bars: 1 cm.*

Fossilization by limonite is rarely documented because this type of fossil is particularly vulnerable to dissolution and/or disintegration. However, it is possible to find fragments of limonitized fossil wood, such as in the glauconitic sands from the Lower Cenomanian of Fouras in Charentes, south-western France. Iron is an abundant component here because the facies combines two sources of it: iron linked to glauconite and iron linked to organic matter. As this highly aerated sand, containing no clay, has been heavily leached, the wood in the deposit became fossilized in the form of limonitic iron hydroxide. However, we have no way of knowing whether that mineralization is original or the result of the transformation of wood that was originally pyritous (see Plate 2.22D).

Hydration of hematite into limonite:

$$Fe_2O_3 + 3\ H_2O \rightarrow 2\ Fe(OH)_3$$

2.4.3. *Siderite nodules*

Sedimentary siderite nodules form in calm, oxygen-poor aquatic environments, either in the deep or very deep ocean, in the brackish supratidal zone, or in a lacustrine environment (Vuillemin et al. 2019; Lin 2020). They are anhydrous iron carbonates that contain no foreign anions ($FeCO_3$). A reducing environment rich in organic matter, in combination with extensive bacterial activity and a high rate of sedimentation in the context of tidal fluctuation, is often favorable to siderite neoformation (Lin 2020). The organic matter from animal or vegetable remains is generally the nucleus of bacterial activity giving rise to siderite. Siderite nodules often oxidize to form limonite, by pseudomorph, when they are exposed to damp air. In France, siderite nodules have yielded excellent fossils in terms of the quality of their preservation in the Carboniferous series at Montceau-les-Mines, Saône-et-Loire, central France, and the formations dating from the Callovian in La Voulte-sur-Rhône in Ardèche, south-eastern France (Charbonnier 2009) (see Plate 2.21B).

Formation of a siderite nodule:

$$Fe^{2+} + 2\ HCO^{3-} \rightarrow FeCO_3 + CO_2 + H_2O$$

Example of sideritic nodules from the Carboniferous found in Montceau-les-Mines, central France

Since the 1970s, the coal quarries in the Montceau-les-Mines sector (Saône-et-Loire, central France) have yielded a multitude of fossils preserved in sideritic concretions formed from clay-rich silts. During the digs conducted between 1979 and 1982, over 120,000 nodules were collected (Perrier and Charbonnier 2014). The paleontological record contained in these ferrous nodules includes a wide variety of freshwater fauna (annelids, crustaceans, myriapods, insects, arachnids, fish, tetrapods) (see Plate 2.21C), and various groups of land-based plants (see Plate 2.21D). The organisms were fossilized with little disarticulation of the bodies, and the soft tissues are preserved in 3D. Very fragile structures, such as the delicate appendices of certain small arthropods, are astonishingly well preserved. Fossilization in the form of ferrous nodules, resistant to diagenesis, limited the compaction of those fossils. Here, such preservation can be understood in a series of stages:

1) The remains were buried in a fine, anoxic mud, which limited decomposition and disarticulation.

2) The anoxia, coupled with a high iron concentration in the environment of the deposit, led to very rapid siderite precipitation (probably due to bacterial activity), and began nodulization.

3) Later phases of secondary mineralization (for example calcite, pyrite) filled the remaining cavities (Perrier and Charbonnier 2014).

2.5. Amberization

Amberization is the transformation of plant resin or sap into amber or copal. The subtle difference between these two materials, of organic origin, which often yield many fossils, are important, but not always entirely agreed upon.

Fossilization in amber or copal, notably in the case of insects, results from their first becoming trapped in flows of scar resin (in the broad sense), produced by various types of wounds in productive trees (see Figure 2.2): on the trunk, branches, leafy twigs, roots, etc. It is even possible for the organisms to become trapped in drops or flows or resin that have fallen to the ground, or in internal fractures in the wood. This is a completely passive process in the case of plant remains, fungi or arthropods, which are caught simply because they were "in the path" of the resin flow. However, it is also possible that certain arthropods were attracted by the aromatic molecules in the resin, and found themselves stuck after coming to the site deliberately. Finally, the wind or spray may have carried various organisms or debris onto the resinous outpourings (acarids, leaves, feathers, hairs, aerial or aquatic micro-organisms, etc.).

Figure 2.6. *Formation of fossiliferous amber*

NOTE ON FIGURE 2.6.– *Diagram indicating the various points at which flows of resin (sensu lato) may have stuck arthropods, or any other organism present on the trees, on the ground, or carried there by the wind. A: flows from growth fissures on the trunk where new branches erupt, or at the nodes of scar tissue resulting from falling branches; B: flows from cuts or breaks in the trunk or branches; C: outpourings from internal fractures in the wood; D: flows where the roots emerge from the trunk; E: outflows from the roots (including in the ground or in water); F: drops or flows of resin that have fallen to the ground (litter or tide wrack); G: cross-section of twigs or branches that have fallen to the ground (figure adapted from Perrichot (2005)).*

2.5.1. *Copal*

Certain authors presented an initial definition of copal, and the crux of its difference from amber, in the fact that it was fossilized angiosperm gum rather than fossilized conifer resin. However, a more commonly used definition refers not to the botanical origin of the preserved plant fluid but to its age, and consequently the degree of chemical transformation which it has undergone. Thus, copal is considered to be a fossilized resin or gum from tens or hundreds of thousands of years ago, or in exceptional circumstances, around one million years ago. Therefore, copal can be considered a subfossil, at least in its Holocene varieties. With respect to its original chemical composition, this fluid has undergone polymerization causing it to harden whilst losing some but not all of its aromatic molecules. In addition, unlike amber, copal is often soluble in alcohol.

The most famous of the copals valued by paleontologists and collectors are Madagascan copal and Colombian copal (Penney and Preziosi 2010). Madagascan copal was produced by angiosperms (flowering trees of the type *Hymenea*). The date attached to it varies widely from one site to another, and indeed from one author to another. The oldest estimates put it at 4 or 5 million years old; the most conservative put it at only 10,000 to 50,000 years. On average, it is taken to be 1 to 2.5 million years old (Schlüter and von Gnielinski 1987). The age of Colombian copal (from the Santander region) is also hotly debated, averaging around 500,000 years, but with extremes from 2.5 million years to 200,000 years (Ragazzi et al. 2003). Colombian copal often holds rich deposits of termites.

2.5.2. *Amber*

In contrast to copal, an initial definition of amber, in the view of certain authors, was that it was fossilized conifer resin, rather than angiosperm gum. The generally held definition, however, identifies amber as resin or gum which was fossilized several million years ago. We find ambers dating from the Cenozoic and the Cretaceous (Penney 2010).

It should be noted that ambers earlier than the Cretaceous are extremely rare, and there are scant indications of amber in the Carboniferous (Sargent Bray and Anderson 2009), the Triassic (–252 to –201 Ma) (Roghi et al. 2006) and the Jurassic (Azar et al. 2010), in contrast to the hundreds of sites the world over dating from the Cretaceous (Penney 2010). In terms of the original chemical composition of the plant fluid, it has undergone a far-reaching polymerization, hardening it and making it insoluble in most varieties, and its has lost all of its aromatic molecules, making it completely odorless after breaking or friction.

Two geological periods, marked by hot climate peaks, were particularly productive in terms of amber: in the Mesozoic, the Mid-Cretaceous, between the Aptian (–121 to 113 Ma) and the Turonian (–94 to –90 Ma, and, more specifically, the Cenomanian, –100 to –94 Ma); then, in the Cenozoic, the Paleogene, and above all the Ypresian (–56 to –48 Ma) – the lower part of the Eocene.

Ambers dating from the mid-Cretaceous are known practically all over the world, especially in Spain, France, Lebanon and Burma (Penney 2010). Most are associated with fossilized wood of the type *Agathoxylon*, often attached to Araucariaceae but also, to a lesser extent, wood from trees of the type Brachyoxylon, belonging to the extinct family of Cheirolepidiaceae. Thus, these two families of conifers are most often considered to have produced the resin in the mid-Cretaceous which went on to become amber (Perrichot 2005).

Ambers from the Paleogene are primarily known in Europe: in France, Poland and in various former Soviet Union countries (Penney 2010). While the Paleogene ambers from northern, eastern and central Europe are attributed to conifers, such as the iconic Baltic amber, associated with characteristic fossilized wood of the pine family, the Eocene amber from France, largely found in Oise, was mainly produced by angiosperms similar to Caesalpiniaceae, whose fossilized wood is *Aulacoxylon* (Nel et al. 1999; De Franceschi and De Ploëg 2003).

Example of inclusions in Cretaceous amber from the Charentes, south-western France

The lignitous deposits from the Albian-Cenomanian found in Charente and Charente-Maritime, south-western France, locally contain a large quantity of fossil resin. Amber from the Charentes has the peculiarity of being highly fossiliferous, having yielded a multitude of arthropods (Néraudeau et al. 2002; Perrichot 2005) and micro-organisms (Breton and Tostain 2005; Girard et al. 2009). In much rarer cases, it has given us dinosaur feathers (see Plate 2.23A and 2.23B), mammal hairs (see Plate 2.23C) and fragments of shed skin from squamata reptiles (see Plate 2.23D) (Perrichot et al. 2008; Vullo et al. 2010).

Generally speaking, insects are the fossils most commonly found in this Cretaceous amber – particularly flies and mosquitoes (diptera, see Plate 2.24B and 2.24C), wasps and ants (hymenoptera), crickets (orthoptera, see Plate 2.24A), cockroaches (see Plate 2.24D) and shield bugs (heteroptera, see Plate 2.24E). In the majority of cases, they are limited to the animal's external shell, with the internal structures and fluids not being preserved.

However, in rarer cases (example of specimens from the Lower Cretaceous in Spain) (Soriano et al. 2010), the animal's internal organs have been preserved. Curiously, plants are among the remains that are fairly uncommon in the Cretaceous amber from the Charentes, and are largely limited to branches of conifers (of the genera *Pagiophyllum* and *Glenrosa*) (see Plates 2.25A, 2.25D, 2.25E and 2.25F) and their cones (of the genus *Classostrobus*) (see Plates 2.25B and 2.25C).

In 3D imaging analysis (see Box 2.7), rare known specimens of conifers have shown that in addition to preserving the cuticle, in certain cases, these plant inclusions have all the internal tissues preserved, right down to the cellular scale (Moreau et al. 2017b) (see Plate 2.25E and 2.25F). Such cases of exceptional preservation can be attributed to the dehydration of the tissues, accompanied by a permineralization of the intracellular spaces by resin. The original composition of the tissues is generally lost, making it unlikely that fossil DNA can be extracted.

When the amber is clear, the fossils it contains can be studied by "traditional" microscopy techniques. However, one of the defining characteristics of Charentese amber is that it is sometimes opaque. Thus, it becomes impossible to study that resin through transparency. *Synchrotron X-ray microtomography* has proved to be a high-performance imaging technique for 3D study of fossilized inclusions preserved in opaque amber (Tafforeau et al. 2006; Lak et al. 2008; Soriano et al. 2010). It enables us to take virtual slices and 3D dissections of the blocks of amber and the specimens that they contain. Thus, synchrotron tomography has revealed internal structures, which cannot be seen from the outside, non-destructively. Above all, this technique enables us to work with very high image resolution (far greater than that of medical scanners, for example). Using this technique, the Grenoble synchrotron (European Synchrotron Radiation Facility) has been able to image internal organs of insects and plant cells in opaque amber from the Charentes (Soriano et al. 2010; Moreau et al. 2017b) (see Plate 2.25E and 2.25F).

Box 2.7. *Microtomography – a 3D imaging technique which reveals fossilized inclusions that are "hidden" in Charentese amber*

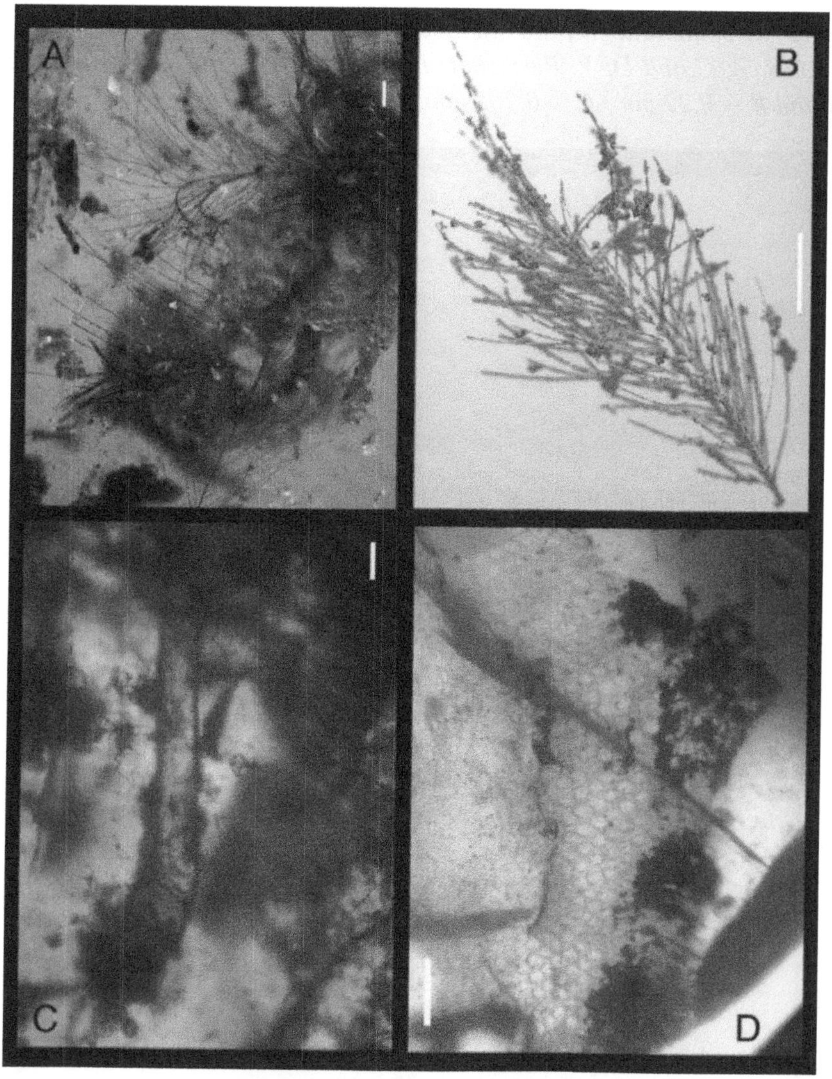

Plate 2.23. *Phanera preserved in amber*

NOTE ON PLATE 2.23.– *Fossils of phanera in Cenomanian amber from Charente-Maritime, south-western France. A and B: dinosaur microfeathers, under a transparency microscope (A) and viewed with synchrotron microtomography (B); C: mammal hair; D: shed skin from a squamata reptile (collections: Néraudeau-Perrichot-IGR; photos: A, C and D: V. Perrichot; B: P. Tafforeau-ESRF). Scale bars: A and B = 0.22 mm; C = 0.205 mm; D = 1 cm.*

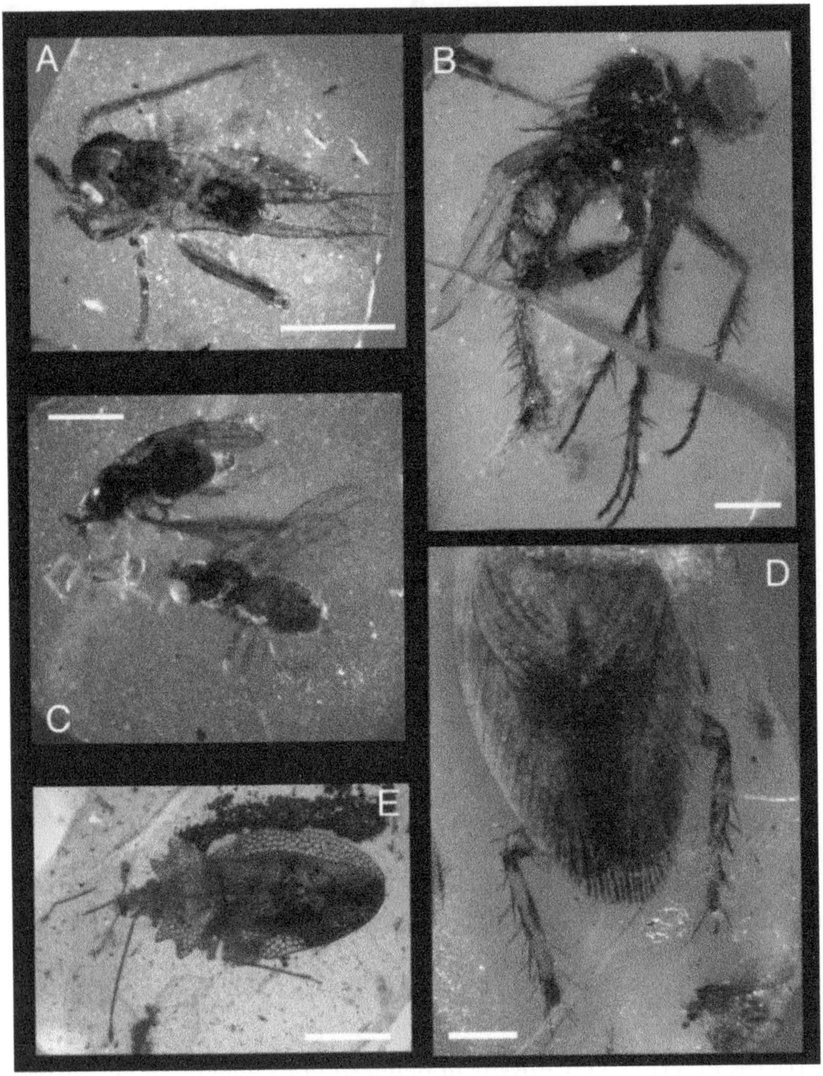

Plate 2.24. *Insects preserved in amber from the late Albian/lower Cenomanian, found in Archingeay-Les Nouillers, Charente-Maritime, south-western France*

NOTE ON PLATE 2.24.– *A: Orthoptera Grilloidea; B: diptera Rhagionidae; C: two diptera Dolichopodidae sharing an acarid; D: cockroach Blattulidae; E: hemiptera Tingidae (collection: Néraudeau-Perrichot-MNHN; photos: V. Perrichot). Scale bars: A and B = 0.25 mm; C to E = 1 mm.*

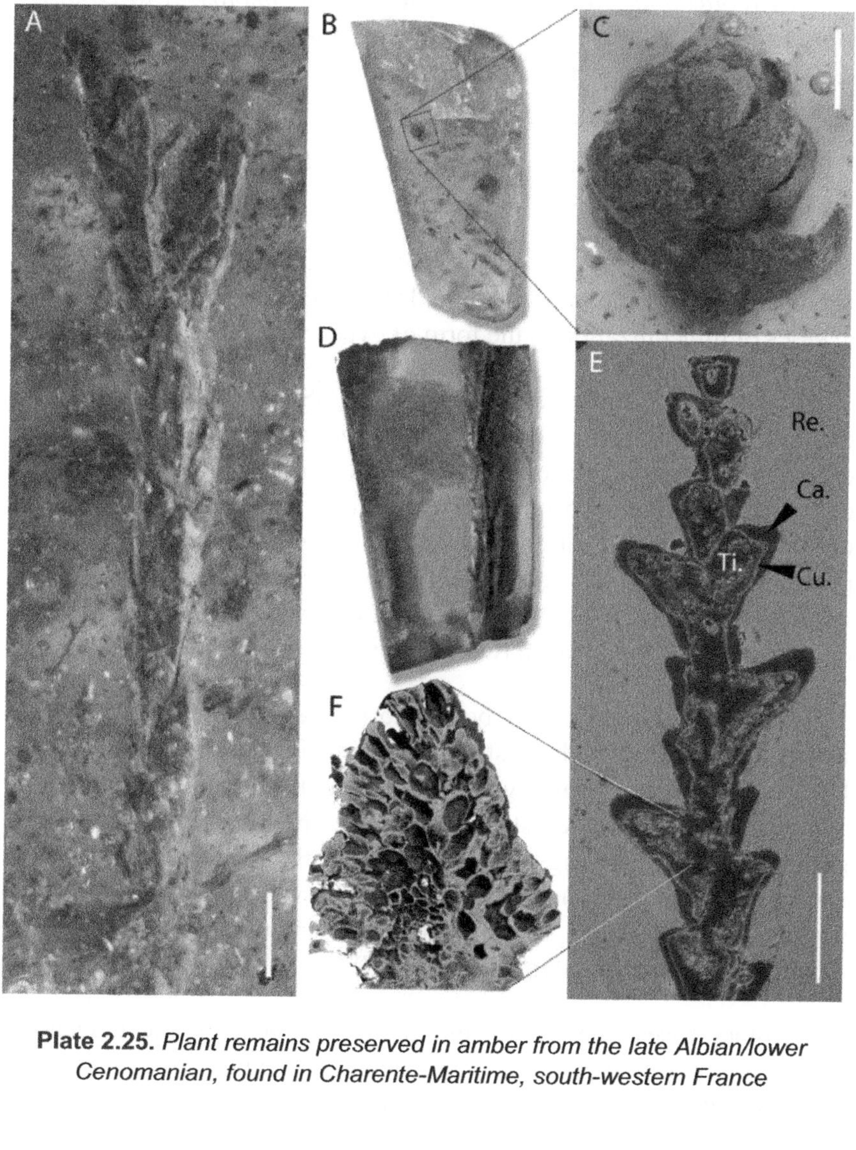

Plate 2.25. *Plant remains preserved in amber from the late Albian/lower Cenomanian, found in Charente-Maritime, south-western France*

NOTE ON PLATE 2.25.– *A: leafy branch of a conifer,* Pagiophyllum*; B and C: wide shot (B) and close-up (C) of a cone; D: leafy branch of a conifer,* Pagiophyllum*; E and F: a virtual cross-section, captured by synchrotron microtomography, of a branch of a conifer,* Glenrosa, *in wide shot (E) and close-up of the internal cellular structures of the leaf (F) (collection: Néraudeau-Perrichot-IGR; photos: A to D: J.-D. Moreau and V. Perrichot; E and F: J.D. Moreau-ESRF). Scale bars: A, B, D and E = 1 mm; C = 0.25 mm.*

2.6. Sulfatization (gypsification and mineralization as barite)

Of all the processes of original mineralization of fossils, sulfatization is probably one of the rarest, and the least spectacular. It is seen mainly in cases of mineralization in the form of gypsum or barite.

Gypsum is a hydrated calcium sulfate ($CaSO_4.2H_2O$) which is very common in sedimentary series, with vast geographic and stratigraphic extensions. It is a rock produced by the evaporation of confined pools of water, and thus gypsums are evaporites from the crystallization of salts – generally in a marine environment. Potentially, their precipitation as crystals could have trapped organic remains present in the lagoon context. However, this process rarely occurs, and where it does occur, it may appear in two different ways.

Firstly, gypsification may be due to the percolation of a fluid rich in calcium sulfate which crystallizes, encrusting mollusk shells in the lagoon environment. Impregnation may go further than this, and in addition to the microcrystals encrusting the shells, gypsum may gradually, and to varying degrees, substitute the carbonate of the shell itself (see Plate 2.26C). Insect wings or larvae included in selenite gypsum (which forms large, translucent crystals) have also been mentioned by certain authors (Tillyard 1922).

In other cases, the cavity left in the sediment by the diagenetic dissolution of a shell based on aragonite may have been filled with fluid enriched with calcium sulfate, and thus created a 3D “mold” or replica of the original shell (see Plate 2.26B), or a replica with gypsification of the shell and its cavities (see Plate 2.26A).

2.6.1. *Gypsification*

Plate 2.26. *Gypsum or barite fossils*

NOTE ON PLATE 2.26.– *A and B: gypsum fossils from Argentina. A: gypsified replica of a bivalve and its internal cavity; B: gypsified replica of the shell of a gastropod, without the internal cavities having been filled; C: gypsified oyster in the porosity of the shell with crystalline outgrowths of gypsum on the surface, dating from the Messinian, found in Los Yesos, Andalucía, Spain; D and E: barite fossils from Steinhardt, Rhineland-Palatinate, Germany (collections: A and B: B. Hogrel; C: Néraudeau-IGR; D and E: D. Gendry; photos: D. Néraudeau). Scale bars: A to C = 5 mm; D and E = 1 cm.*

2.6.2. *Mineralization to barite*

Barytin (or barite) is a barium sulfate with the formula $BaSO_4$, hydrothermal in origin, meaning that it comes from highly mineralized hot-water springs. It is found in a variety of crystalline forms, but usually sprays or piles of flat, white or clear crystals. Its hydrothermal origin means that it is formed in contexts where fluids circulate and can percolate through fossiliferous rocks. When the fossils have internal cavities, the barite may crystallize within them, while the original shell is dissolved. The result therefore is a stack of flat crystals that fills the fossil more or less completely. One very well known barite deposit is the site of Oligocene fossils in Steinhardt, Germany (see Plates 2.26D and 2.26E).

In western France, the Hermenault site in Vendée, dating from the Lower Jurassic, contains large detritic facies in which fossilized plants, notably hollow casts of the stems of equisetales ("cousins" to horsetails), are coated with barite (Diot et al. 2007). At Payré Point (Jard-sur-Mer), again in Vendée, calcitic shells of bivalves dating from the Lower Jurassic have been epigenized to barite by hydrothermal solutions. The geological context lent itself to multiple hydrothermal mineralizations by the circulation of fluids at the junction between the bedrock and the sedimentary Liasic upper strata (notably metal hydroxides and sulfides, and silicification) (Pouclet 2019).

2.7. Phosphatization

2.7.1. *Phosphateous nodules*

Nodules are subglobulous concretions of varying nature (carbonaceous, phosphate, ferruginous, siliceous, carbonate, etc.) whose composition differs significantly from that of their surrounding rock (Thoral 1946). They are generally formed in a context rich in organic material, even if only in the location in which the nodule arises. The bacterial activity by which this organic matter is degraded leads to heterogeneity of the sediment, leading to early diagenesis, and it is that portion of modified sediment which makes up the forming nodule (Bréhéret et al. 2004). Often, the source of organic material is a macro-organism (animal or vegetable) in the process of controlled decomposition; a portion of the bacteria that are involved in its degradation can, at the same time, contribute to its mineralization. In other cases, the nucleus of the nodule is purely bacterial.

Phosphate nodules result from the production of ammonia (NH_3) and tricalcium phosphate ($Ca_3(PO_4)_2$) by the decomposition of organisms (Collet 1906), with these two chemical compounds reacting to produce ammonium phosphate ($PO_4(NH_4)_3$). The action of the ammonium phosphate of the calcium carbonate ($CaCO_3$) in the shells (particularly in brachiopods, crustaceans and mollusks) is the starting point of phosphate nodulization. The pseudomorph of a shell into calcium phosphate ($(PO_4)_2Ca_3$) then constitutes the nucleus (the "seed") of precipitation of calcium phosphates due to reactions between the ammonium phosphate and the calcium bicarbonate in seawater. Glauconia (iron- and potassium hydrosilicate), which is typical of shallow marine, reducing environments, is generally found alongside phosphate concretions. Just like the phosphate nodules, it is characteristic of a slowing of sedimentation, and accompanies those nodules at so-called "condensed" levels. At an oceanographic level, phosphate concretions are believed to generally form where warm waters meet cold, as the thermal shock causes the death of numerous aquatic organisms, which then decompose en masse (Collet 1906; Morozewicz 1924).

A particular process of phosphatization of organisms in subnodular form is more clearly bacterial, notably linked to the activity of mineralizing micro-organisms contained in microscopic animal excrement (pellets) or macroscopic dung (coprolites). It is this excrement which accounts for the majority of the volume of a nodule. The process is particularly characteristic of the excrement of carnivores that have ingested vertebrates (for example fish, or small mammals), or phosphateous debris (teeth, bones, phanera). These elements of animal origin are major sources of phosphate, which, once they have undergone the processes of dissolution and reprecipitation, go hand in hand with bacterial activity (Dornbos 2010), concentrating in coprolites.

Formation of a phosphate nodule:

$$2\ PO_4(NH_4)_3 + 3\ CaCO_3 \rightarrow 3\ CO_3(NH_4)_2 + (PO_4)_2Ca_3$$

2.7.2. *Phosphorites*

In Causses du Quercy (Lot, south-western France), karst depressions formed in the early Tertiary were filled during the Eocene and Oligocene with clays rich in phosphate nodules, before being covered by layers of limestone in the Miocene (–23 to –5 Ma). In these "phosphatous caves" in Quercy, the nodules are composed of phosphorite [3 $Ca_3(PO_4)_2$, $Ca(OH,F,Cl)_2$] – a calcium phosphate with hydroxyl, chloride and fluoride groups. Having been mined since 1870 for phosphates to use in fertilizer, the phosphatous caves soon proved to contain rich deposits of fossils of vertebrates dating from the Eocene and Oligocene. Hundreds of different species have been recorded there – mainly mammals (over 400 species), but also birds, reptiles, insects and plants (Legendre et al. 1997).

There are three potential origins for the diffuse or nodular phosphorus found in the clay:

1) resulting from the dissolution of the overlying sedimentary layer;

2) resulting from phosphates contained in the bones of vertebrates which fell into the karst depressions;

3) resulting from the excrement ("guano") of bats which lived in the karst network before it was filled.

However, the most recent studies seem to indicate that the phosphorus came from the alteration of deposits from the Upper Cretaceous (Pélissié et al. 2006). Indeed, varying by location, the Upper Cretaceous sediments in western Europe may contain between 1 and 20% phosphorus pentoxide P_2O_5. The study of around 200 phosphate caves allows us to document the regional ecosystems in the Eocene and Oligocene, with 100 different fauna covering a total duration of 30 million years between the oldest and the newest (Mazin et al. 2013). The bones are very well preserved, with the phosphates of the bones being preserved in diagenetic phosphate. Some exceptional fossils have even been found, such as the scaly skins of reptiles (see Plate 2.27A), and an almost mummified frog, complete with its skin and eyes in 3D (Lemierres et al. 2021) (see Plate 2.27B).

2.7.3. *Crystalline phosphate fossils*

There are examples of non-nodular phosphatizations of fossils, generally in the form of crystals that have developed in the internal cavities of shells. In the Kerch Peninsula in Crimea (Ukraine), we have seen crystallizations of anapaite ($Ca_2Fe(PO_4)_2.4H_2O$) in fossils of bivalves. This light green-colored mineral is a complex phosphate compound of calcium and iron. Similarly, crystals of vivianite, a hydrous iron phosphate ($Fe_3(PO_4)_2.8(H_2O)$), are present in the inter-valve cavities of fossilized bivalves in the same region, in the form of green/dark blue needlelike formations (see Plate 2.27C and 2.27D). Such crystallizations of vivianite have also been found in or on fossils of bones dating from the Ancient Quaternary – notably the tusks of proboscideans from Guatemala and Idaho (Gaubert 1904; Shannon 1921). They are also fairly common on bones from the Holocene, found during archeological digs. Belemnites from the Cretaceous with pseudomorph into vivianite have been found in the region of New Jersey (Mullica Hill), USA (Weller 1907).

In addition to vivianite, crystals of ludlamite, a hydrous iron phosphate including varying quantities of manganese and magnesium ($(Fe, Mg, Mn)_3(PO_4)_2.4H_2O$), have been known to develop at the edges and on the surface of certain fossils, such as on the gastropods found in Cabeça do Cachorro (São Gabriel da Cachoeira), in Brazil (see Plates 2.27E and 2.27F) (Porpino Lemos et al. 2007).

Plate 2.27. *Various phosphatized fossils*

NOTE ON PLATE 2.27.– *A and B: examples of fossils showing exceptional preservation from the phosphorites formation in Quercy, south-western France. A: snake* Palaelaphis antiquus, *with its scaly skin finely preserved; B: frog* Thomastosaurus gezei, *preserved as a 3D "mummy" with its skin intact. C to F: fossils with phosphate-rich mineralizations. C and D: two pictures of the same bivalve fossilized in vivianite in the Kerch region, Ukraine; E and F: two pictures of the same gastropod with crystalline ludlamite outgrowths, found in São Gabriel da Cachoeira, Brazil (collections: A and B: MNHN, Paris [A = QU17279; B = QU16338]; C and D: É. Dépré; E and F: J.-C. Le Sager; photos: A and B: N. Robin; C and D: D. Néraudeau; E and F: M. Guillo). Scale bars: A to D = 1 cm; E and F = 0.5 cm.*

2.8. Bituminous facies

The formation of fossils partially or completely impregnated with or composed of bitumen may stem from various phenomena. It may be caused by primary "bitumization", in which the hydrocarbons are part of the original context of fossilization, or by secondary bitumization, where the hydrocarbons are injected into a fossiliferous rock, long after the original fossilization occurred.

Example of the bituminous shale in Messel

Primary bituminous facies, and notably shale of this type, are rocks that formed in extremely low-oxygen contexts, containing a high level of organic matter. One of the most extraordinary fossiliferous sites, at Messel in Germany, is known for the process of physical and chemical protection of the fossils found, which was developed by Oskar Kuhn (Kuhn 1962) to save them from disintegrating when the organic matter came into contact with air. The sediment is replaced with resin, and the brown or blackish fossils are exhibited in a yellowish resin medium (see Plate 2.28A). This deposit dates from the Eocene, and its formation is thought to be linked to the eruption of CO_2 and H_2S accumulated in a lacustrine context and released by volcanic activity. This gaseous pollution would have killed airborne species (insects, chiroptera, birds) and land-dwelling species (mammals, reptiles, amphibians, arthropods, plants) near to the lake, and when their remains fell into the anoxic waters of the lake, they were shielded from putrefaction. We also find aquatic lifeforms (for example fish) which lived in the lake and underwent the chemical alteration of their living environment (Smith et al. 2018).

Example of the bituminous limestone in Limagne (central France)

Less spectacular than Messel, but still worthy of note, in Limagne in central France is the Mine des Rois, dating from the Oligocene, in Dallet (Puy-de-Dôme). This site is made up of bituminous limestone, which has preserved a wide variety of fossils of land-dwelling organisms. The bitumen, formed from marl rich in organic matter in the depths of the sedimentary basin, was brought back to the surface through faults or volcanic vents, impregnating the overlying layers of porous limestone (Thomas 2008). The invertebrate fossils found are mainly the larvae of aquatic insects (phryganea) and snails of the genus *Helix* (see Plate 2.28B). The site has also yielded fossils of vertebrates (mammals, birds, reptiles) and eggshells.

Example of the "Schistes Cartons" in the Grands Causses, south-eastern France

In the Grands Causses sector, south-eastern France, the base of the Toarcian (–184 to –175 Ma) is characterized by clay-rich limestone, organized into thin plates, and banks of limestone that are tens of meters thick, with bituminous strata. These deposits, named "Schistes Cartons" for their paper-like texture, can easily be recognized on the edge of the high limestone plateaus in that region because they usually form sheer cliffs, clearly breaking away from the marl deposits lying beneath (dating from the Domerian) and above (from the mid-Toarcian), which are more susceptible to erosion. Schistes Cartons offer up remarkably well preserved, abundant marine fauna. The fossils most often found are ammonites (for example *Dactylioceras*, *Harpoceras*). They are highly compacted (see Plate 2.29B), and sometimes form veritable accumulations in which *Harpoceras falciferum* can reach sizes of over thirty centimeters. The

Schistes Cartons in the Grands Causses also provide fossils of numerous small fish which can be assigned to *Leptolepis coryphaenoides* (Coquel-Poussy 2013) and, more rarely, crustaceans (for example lobsters) (Audo et al. 2017) or the remains of coleoid cephalopods (Jattiot et al. 2024). Sometimes, they even still have their ink sac (see Plate 2.29A). The most spectacular specimens include large ichthyosaurs, found in Hérault, southern France, with their soft tissues partially preserved in 3D (Bomou et al. 2021).

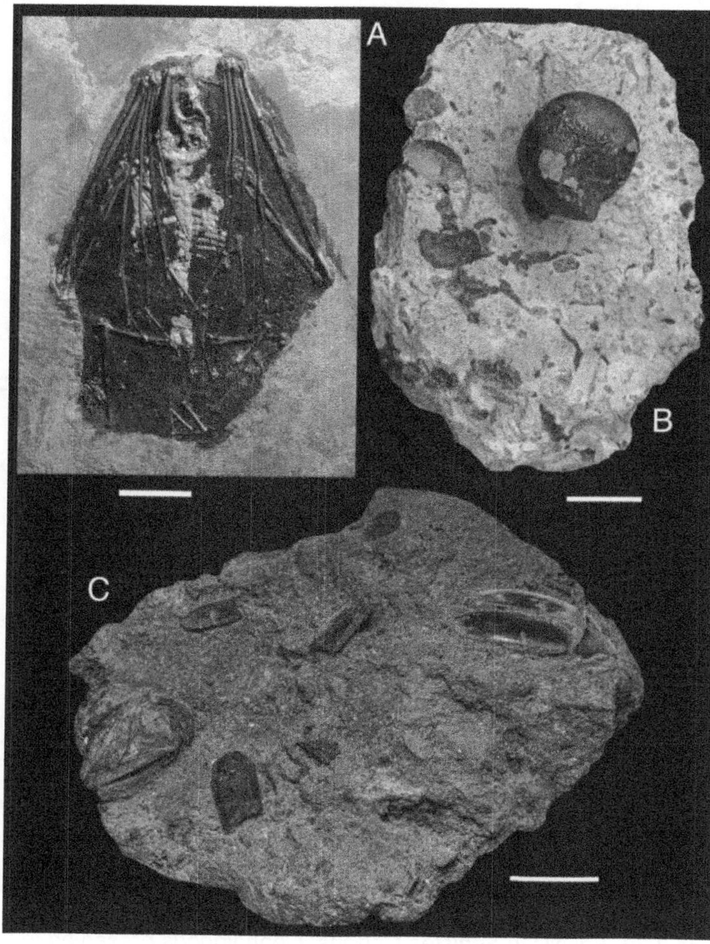

Plate 2.28. *Bituminous fossils*

NOTE ON PLATE 2.28.– *A: chiroptera from the Eocene in Messel, Germany; B: pulmonated snail* Helix ramondi *from the Oligocene, found in Dallet, Puy-de-Dôme, central France; C: coleoptera from the Pleistocene found in the La Brea tar pits, USA (collections: A: Muséum d'Histoire naturelle de Bourges; B: IGR; C: J.-C. Le Sager; photos: A: A. Radix; B: D. Néraudeau; C: M. Guillo). Scale bars: 1 cm.*

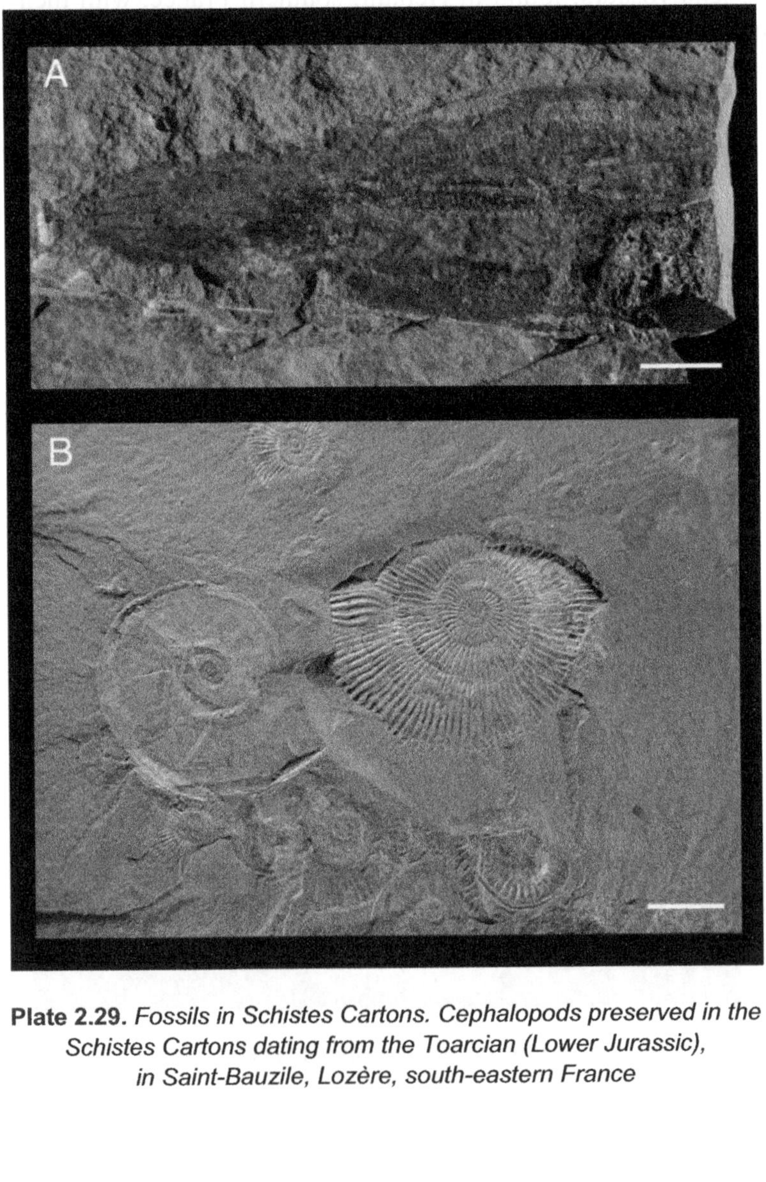

Plate 2.29. *Fossils in Schistes Cartons. Cephalopods preserved in the Schistes Cartons dating from the Toarcian (Lower Jurassic), in Saint-Bauzile, Lozère, south-eastern France*

NOTE ON PLATE 2.29.– *A: coleoid* Loligosepia *sp., with preservation of the bodily cast in the form of reddish iron oxides and the ink sac (right) in a blackish carbonized form; B: compacted ammonites (the largest is a* Dactylioceras *sp.), corresponding to 2D compressions of the shells (collection: N. Coquel-Poussy; photos: J.-D. Moreau). Scale bars: 1 cm.*

During the Basal Toarcian, the Causses region, south-eastern France, experienced a massive rise in sea level, but also an oceanic anoxic event known as the TOAE. Owing to the sharp decrease in oxidation of organic matter, such matter was abundantly preserved in Schistes Cartons, which accounts for the tar-like smell that is characteristic of the rocks in that formation). For this reason, they attracted a great deal of interest in hydrocarbon extraction, such as "shale gas". Note that during the first half of the 20th Century, the tar present in the Schistes Cartons was even extracted for exploitation temporarily in Aveyron, south-eastern France (Defaut et al. 1990).

Example of the La Brea tar pits (USA)

Discovered in 1769, the La Brea tar pits in Los Angeles (USA), were formed over 40,000 years ago in a coastal glacial context. The tar has perfectly preserved the organisms stuck in it by protecting them from oxidative action, whilst also impregnating them with carbon. Around 650 species of animals and plants living in the glacial age are exceptionally well preserved; to date, hundreds of thousands of specimens have been collected (Stock 1992). The vertebrates are the most spectacular examples, but there are also insects present – notably, coleoptera whose elytra are still shiny (see Plate 2.28C).

3

Remarkable Cases and Fossiliferous Contexts

The different qualities of fossilization (internal casts, replicas, hollow impressions, etc.) and the numerous possibilities for mineralization (carbonates, silicates, phosphates, ferrugination, etc.) presented in the previous two chapters do not cover all remarkable fossils and fossiliferous facies. There are also the cumulative contexts which lead to rich accumulations of remains (teeth, bones), sedimentary surfaces which preserve the traces of movement (vermiform trails, footprints, etc.), or rapid burials, which preserve skeletal remains with their anatomical connections intact.

In parallel to the cases mentioned above, fossils can be remarkable because they provide an exceptional picture of the activities (bites, parasitism, etc.) or the productions (excrement, regurgitations, egg-laying) of an organism, which are indicative of the trophic network and the biological functions of extinct species. Finally, contexts of fossilization which are basically unclassifiable are those which lead to

---

For a color version of all the figures in this chapter, see www.iste.co.uk/neraudeau/fossilization.zip.

the mummification of bodies, combining the loss of fluids with the desiccation of the bodily shells, or the organs.

3.1. Burial in fine, layered mud

The burial of organic remains in mud with an extremely fine grain size quickly protects those remains from the action of oxygen and decomposers, even in the finest anatomical details, because the fineness of the sediment means it is able to seal off the smallest interstitial spaces. The repeated deposition of fine sediment on tidal, seasonal or annual cycles contributes to the stacking up of laminated fossiliferous layers. The fine sediment involved may be silt-clay, marl or limestone.

In addition to the beneficial fineness of the sediment, exceptional preservation in fine marine sediments is generally linked to the protective and mineralizing action of microbial mats (Gall 1990, 1995). This action is more difficult to show in lacustrine or paralic fine sediments.

3.1.1. *The role of microbial mats in Precambrian and Cambrian shales*

The development of the metazoans began in the late Precambrian, mainly in the Ediacaran, around 635 to 539 million years ago. Thus, the first marine ecosystems contained only a low-density population of those metazoans, generally with no skeleton, and with an epibenthic life. Decomposers were not yet as abundant as they are in modern ecosystems, and there was very little or no bioturbation of aquatic substrates.

In this very specific context, poor in primary consumers, the microbial mats which developed on the substrates were frequently fossilized. Although they were extremely thin, they were shaped by oscillations in sea currents, taking on finely rumpled or more heavily crumpled morphologies, and sometimes even blistered or dotted (see

Plate 3.1D). At a microscopic scale, the trans-sectional view of these microbial veils reveals their filamentary structure (see Plate 3.1E) (Gougeon et al. 2018).

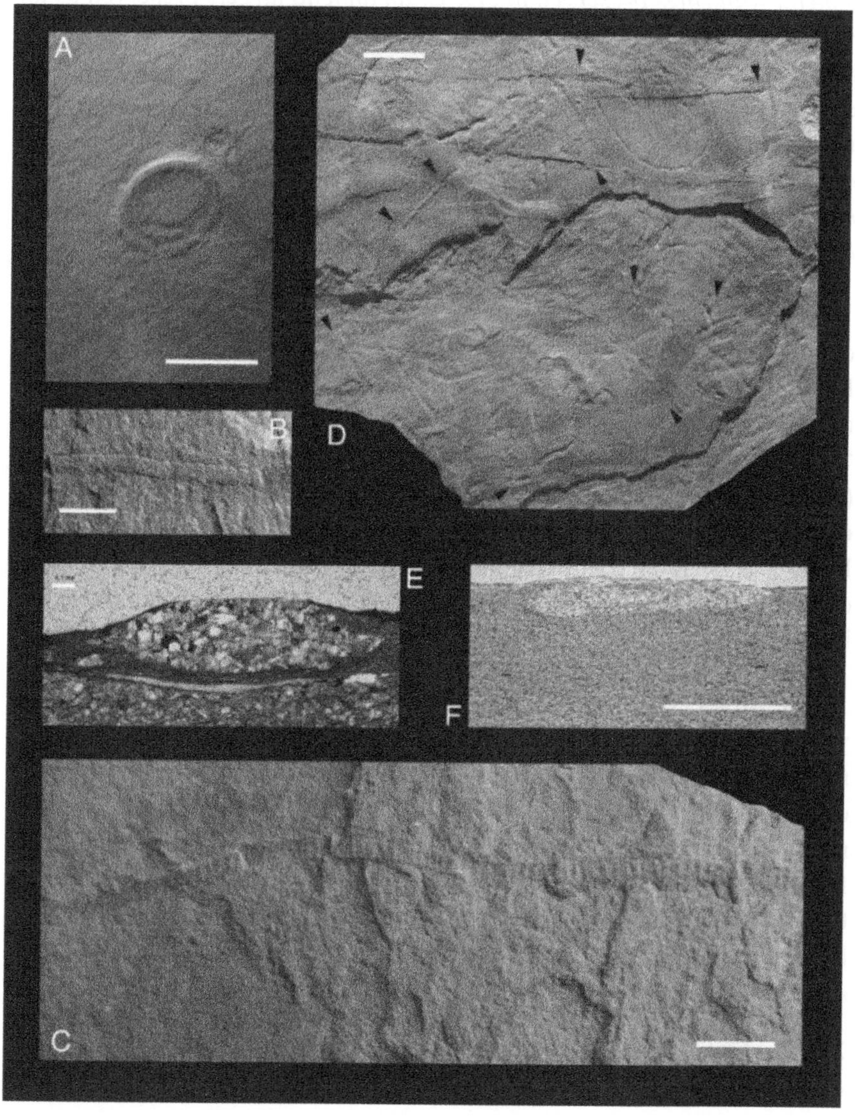

Plate 3.1. *Fossils dating from the late Ediacaran (the "Brioverian") and the Lower Cambrian (Fortunian), found on the Armorican Massif*

NOTE ON PLATE 3.1.– *A to E: Brioverian specimens found in Ille-et-Vilaine (north-western France); F: Fortunian specimen, Cotentin, Manche (north-western France). A:* Nimbia-*like medusoid; B: an indeterminate annelid with rings preserved in 3D; C: an annelid largely preserved as an imprint; D: bacterial mat of "Neantia" structure, traversed by burrower trails made by annelids (indicated by arrows); E: thin film showing the cross-section of an annelid on a bacterial mat, the latter appearing dark with interwoven blackish filaments; F: thin film showing the cross-section of a medusoid deposited on the sediment and covered by a brownish-yellow bacterial mat (collection: Néraudeau-IGR; photos: D. Néraudeau). Scale bars: A and D = 1 cm; B, C and F = 5 mm; E = 0.1 mm.*

In addition to their own exceptional fossilization, these mats covered the bodies of macro-organisms present on the sea bed (see Plate 3.1F) and facilitated the remarkable preservation of various metazoans with no internal or external skeletal reinforcements, such as "medusoids" (see Plate 3.1A) or annelids (see Plates 3.1B and 3.1C) (Néraudeau et al. 2018). These crumpled bacterial mats have even often been fossilized with the trails of annelids who fed on them (see Plate 3.1D). Indeed, microbial mats represented a resource for the earliest grazers, and we have found fossils of bacterial mats traversed by the trails (feeding networks) of vermiform animals (annelids).

The first ecosystems of the Cambrian, around –539 to –530 million years, as yet had a low density and bioturbations and populations of metazoans, and although that population included increasing numbers of skeletized organisms, they did not completely destroy the bacterial mats. Fossilized microbial mats are indeed present in the deposits from the early Cambrian (see Plate 3.1F) (Néraudeau et al. 2023), but they became rarer in more recent layers as they began being consumed by grazing organisms, which were more numerous by then.

3.1.2. *The role of microbial or algal mats in pelites and lithographic limestones*

Microbial mats (algal or bacterial mats) play major roles in the exceptional fossilization of metazoans which, in the absence of these micro-organisms, are only ever poorly fossilized, if at all: they stabilize the substrate, form a barrier between the dead organism and physical attack (erosion, dislocation) or chemical attack (oxidation) and may contribute to the mineralization of the most fragile tissues. Such is the case with the phanera and soft tissues in vertebrates, and especially in the case of invertebrates, which do not contain a mineral structure (jellyfish, nudibranchs, etc.) (Gall 1990).

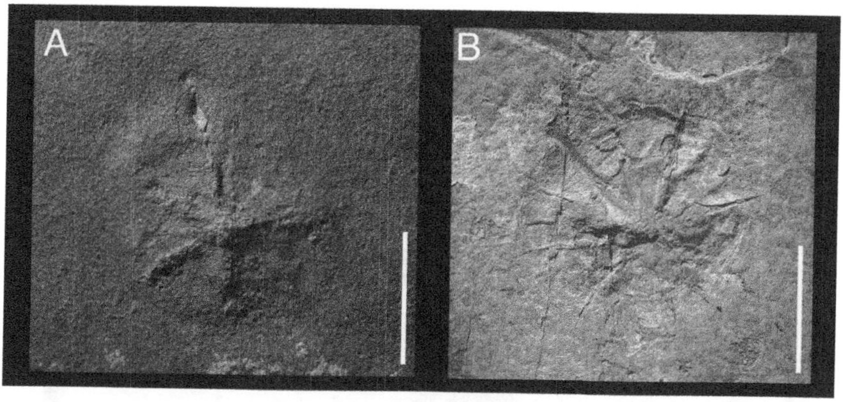

Plate 3.2. *Medusoids preserved in laminites*

NOTE ON PLATE 3.2.– *A: cnidaria* Medusina atava, *fossilized in laminites from the Permian of Aveyron (south-eastern France); B: indeterminate medusoid found in laminites dating from the Oligocene, found in Lubéron, south-eastern France (collections: A: Musée de Lodève; B: Lubéron geological reserve; photos: A: J.-D. Moreau; B: D. Néraudeau). Scale bars: A = 3 mm; B = 3 cm.*

Example of Permian pelites from the south of the Massif Central

The Permian basins surrounding the Grands Causses region (south-eastern France) preserve thick sedimentary series formed in a continental context. A portion of the wine-colored pelites which crop out in the basins of Lodève, Saint-Affrique and Rodez formed in a playa-type environment (Lopez et al. 2008). These vast plains, fed by water courses, were periodically occupied by temporary lakes. The alternation of repeated periods of emersion and immersion, combined with a fine sedimentation and the presence of microbial mats developing in fine films of water, allowed the fossilization of an abundance of ichnofossils (the tracks of amniote and anamniote tetrapods, traces of the swimming of fish, trails and burrows of various protostomia), arthropods (insects, conchostracan and triopsid crustaceans) or freshwater jellyfish (Ellenberger 1984; Gand et al. 1996, 1997, 2000; Béthoux 2008; Moreau et al. 2024) (see Plate 3.2A).

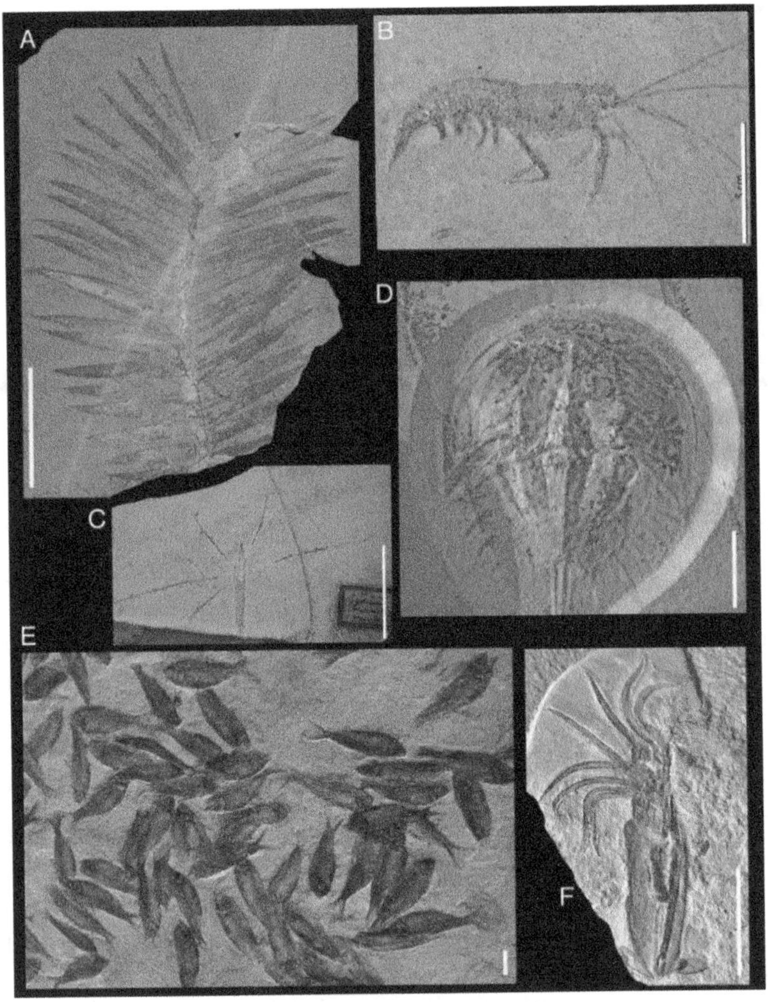

Plate 3.3. *Fossils in lithographic limestone*

NOTE ON PLATE 3.3.– *A: frond of benettitalean* Zamites *from the Kimmeridgian of Nivoliers, Causse Méjean, Lozère (south-eastern France); B: decapod crustacean, dating from the Kimmeridgian, found in Solnhofen, Bavaria, Germany; C: insect from the Kimmeridgian, found in Solnhofen, Bavaria, Germany; D: limule* Mesolimulus *from the Kimmeridgian of Solnhofen, Bavaria, Germany; E: fish* Knightia *from the Eocene of Wyoming (Green River Formation), USA; F: cephalopod (squid) from the Cenomanian of Lebanon (collections: A: G. Lafaurie; B and C: Université de Bourgogne; D and E: É. Dépré; F: M. Habert – Musée Vert, Le Mans; photos: A to C: J.-D. Moreau; D and E: D. Néraudeau; F: N. Morel). Scale bars: 5 cm.*

Example of lithographic limestone from the late Jurassic

The Late Jurassic – particularly the Kimmeridgian geological stage (–157 to –159 Ma) and the early Tithonian (–149 to –145 Ma) – was a period marked by paleoenvironments of shallow lagoons (brack), which developed in protected areas, notably behind barrier coral reefs. These protected areas saw the deposition of very find carbonated sediments, with weak hydrodynamism, and constituted receptacles for the remains of organisms from both the marine environment (see Plates 3.3B and 3.3D to 3.3F) and the land or rivers. Indeed, the relatively enclosed, oxygen-poor environment, with hot water and fluctuating salinity, caused the mass death of aquatic vertebrates and invertebrates (see Plate 3.3E), whilst receiving the remains of animals or plants carried out from land by estuaries, or by the wind (insects, leaves) (see Plates 3.3A to 3.3C). The fine sediment was deposited in successive layers which covered and impregnated the organic remains with a film of sludge, more or less rich in mineralizing micro-organisms. Due to bacterial action and lack of oxygen, the fragile tissues (skin, phanera, foliar limbs, arthropod cuticles, etc.) have not degraded much and have been mineralized as iron oxide (reddish brown) or manganese oxide (black). The most common fossils are fish, preserved individually or en masse (see Plate 3.3E). However, there are also various echinoderms (sea urchins, starfish, brittle stars, unaffixed crinoids of the comatula group) and bivalve mollusks or cephalopods (see Plate 3.3F). In south-eastern

France, this type of deposit is known in remarkable sites with exceptional fossilization, such as the Lagerstätten in Canjuers (Var), Cerin (Ain), and the Causse Méjean (Lozère). Note that in Lebanon, comparable facies, famed for their fossils of fish, are known in layers from the Upper Cretaceous (Cenomanian and Santonian; for example the Sahel Alma deposit) (Audo et al. 2012; Jattiot et al. 2015b).

Example of Oligocene laminar limestones in Lubéron (south-eastern France)

The Oligocene laminar limestones in Lubéron are finely layered carbonated rocks, formed by the sequential decantation of fine mud, rich in bacterial mats. The layers are sometimes highly fossiliferous, and they have produced remarkable fossils of fish, insects, bird feathers, but also, more surprisingly, a large quantity of "medusoids" (Nel et al. 1987; Gaudant et al. 2018). More specifically, these late fossils are of gelatinous cnidaria living in colonies. Their tissue mass was almost entirely made up of water, with a small proportion of organic matter (see Plate 3.2B). Despite this very low original mineralization of the tissues, the fineness of the sediment, finely impregnating the tissular porosity, and the presence of a bacterial mat covering the dead specimens, protecting them from oxygen and contributing to their mineralization, allowed for the exceptional fossilization of organisms which are rarely preserved elsewhere.

3.1.3. *Anoxia of laminar clays bearing plants*

Example of laminar clays containing Cretaceous plants

In the mid- and Upper Cretaceous, largely at the end of the Albian (–113 to –100 Ma) and more so at the start of the Cenomanian (–100 to –94 Ma), periods with a low sea level, in transgressive contexts, allowed clay to be deposited in the paralic basins. The associated paleoenvironments were temporary lagoons with a variable level of salinity in which clay with an anoxic tendency was deposited by decantation during the phases of isolation from tidal inlets. These paralic basins, which occasionally saw the entry of fish, crustaceans or mollusks from the sea, were more frequently the receptacles of feathers (see Plate 3.4A), plants (see Plates 3.4B and 3.4C) and insects (see Plate 3.4D) from inland, carried there by water courses (Vullo et al. 2013; Néraudeau et al. 2019). The burial of the plant and insect remains in the fine layers of clay, combined with anoxic mud and deep water, doubly protected the remains from putrefaction linked to oxygen and decomposers, whilst impregnating, or encrusting the finest details of the surfaces of the plant and animal remains. Thus, leaves are fossilized as impressions, showing the details of their outlines, their petiole, their nervation with a brownish tinge, which indicates a residue of organic matter (see Plate 3.4B). In certain clay facies, the fossilization of plants is not limited to such organic residue, having preserved a compression of the cuticle in the finest of details (imprints of epidermal cells, stomata, pilosity, etc.) (see Plate 2.12 and Figure 2.2) (see section 2.2).

Example of Oligocene laminar clay, Brittany

During the Oligocene, the Chartres-de-Bretagne basin, near Rennes (north-western France), was marked by the deposition of dark-colored clay corresponding to sapropelic deposits (i.e. anoxic and rich in organic material) accumulated in a freshwater body. In an oxygen-poor aquatic context, the mass mortality of various species led to the accumulation of bodies on the silty seabed, with the fine sediment having both buried and impregnated the most fragile organisms. In this way, these clays fossilized both aquatic plants (nenuphars) and fish (Lemierre et al. 2023).

In this latter case, the fossilization is particularly remarkable in that, in addition to the vertebrae preserved in 3D and the finely preserved scales and "bones", we find brown or black organic matter which indicates the soft tissues – particularly the viscus and eyes (see Plate 3.4E). The slowing and the relatively limited degree of putrefaction, in this context of both lack of oxygen and abundance of diffuse organic matter, even allowed for the fossilization of vertebrates rarely found in the fossil record, such as frogs (see Plate 3.4F).

Plate 3.4. *Fossils in laminar clay*

NOTE ON PLATE 3.4.– *A to C: fossils in Cenomanian clay, Tonnay-Charente, Charente-Maritime (south-western France): A: dinosaur feather; B: crenulated leaf; C: flower buds; D: insect wing; E and F: fossils in Oligocene clay, Chartres-de-Bretagne, Ille-et-Vilaine (north-western France): E: fish; F: frog (collections: A, B and D: Dépré-IGR; C: Université de Lyon-É. Dépré; E and F: IGR; photos: A: R. Vullo; B: J.-D. Moreau; C: T. Le Diouron; D: D. Néraudeau; E and F: D. Gendry). Scale bars: 1 cm.*

3.1.4. *Diatomites*

Diatoms are single-celled green algae with a siliceous shell (frustule), which can accumulate over the course of repeated blooms, and then form biogenic deposits known as diatomites. These deposits lead to rocks which may be compact or laminar. When they are laminar, we see an alternating pattern of light- and dark-colored layers, which indicate successive seasonal blooms. Diatomitic blooms are linked to conditions of temperature and eutrophication which can lead to the mass death of other organisms – notably fish or other aquatic vertebrates (see Plates 3.5A, 3.5E and 3.5F), especially as certain diatoms, like other single-celled algae (dinophyceae), produce high levels of toxicity. Other remains of organisms, transported by the wind (insects, leaves, feathers and hairs), may also be deposited in water enriched with diatoms and be found in diatomitic laminae (see Plates 3.5B to 3.5D).

Both in the case of autochthonous aquatic organisms and in that of fossils brought to a secondary site by the wind or water, the fineness of the diatomitic mud and its reducing context allow for exceptional preservation of all the organisms deposited there, including their phanera, their cuticles or their soft tissues. In addition, numerous species of diatoms produce a mucus which forms a biofilm over the substrate, slowing putrefaction whilst also favoring the mineralization of the organisms present on the bottom. The resulting diatoms and diatomites are found both in freshwater and marine environments.

Example of the Miocene diatomites, Saint-Bauzile (Ardèche, south-eastern France)

In a freshwater context, diatomitic deposits are often linked to volcanic activity which provides the dissolved silica (e.g. ash fallout) necessary for diatoms' development. Thus, in Ardèche (south-eastern France), diatomites from the Upper Miocene (7.5 Ma) from the basalt plateau of Coiron, extracted from the Saint-Bauzile quarry, formed in a crater lake (maar) 2 km in diameter. Diatomites, deposited on marls from the Lower Cretaceous (Valanginian), were covered by lava dating from the Messinian (–7.2 to –5.3 Ma). When studied (Iskandar 1990), they revealed that the diatomitic laminae were lain down over the course of 400,000 years, with annual varves. This cycle of deposition with a very fine temporal resolution saw a rich succession of fossiliferous layers containing a broad range of aquatic and terrestrial organisms, such as fish, insects, foliage, and also pieces of skin and plumage of vertebrates (Lebrun 2013; Saint-Martin et al. 2017) (see Plates 3.5A to 3.5E).

Example of the Messinian diatomites in the south-eastern Mediterranean

Also in the Upper Miocene (the late Messinian, around 5.5 to 6 Ma), but in the marine context, a process of drying of the Mediterranean altered the physico-chemical characteristics (salinity, pH, oxygenation) of the sea. Certain basins were isolated from the main basin by the dropping water levels, notably in Italy, Algeria (Oran region) and south-eastern Spain. Such small, confined basins were advantageous for the proliferation of certain single-celled algae, such as diatoms, which thrive particularly in destabilized aquatic environments. Their proliferation, along with their mass accumulation on the substrate, contributed to the consumption of the resources in the environment and, consequently, the death of other organisms – notably fish. Thus, diatom blooms caused the mass death of small fish, which accumulated on the bed already enriched with these microalgae. Due to lack of oxygen and decomposers, and being quickly buried by films of diatoms, these fish were preserved in massive numbers in the laminae of diatomites – rocks formed by the accumulations of diatoms. The fossils of these aquatic vertebrates were not only preserved with all their skeletal elements and connective tissues, but also often with the brownish print of their bodily envelope and of their tissues, imprinted in the diatomitic laminae (see Plate 3.5F).

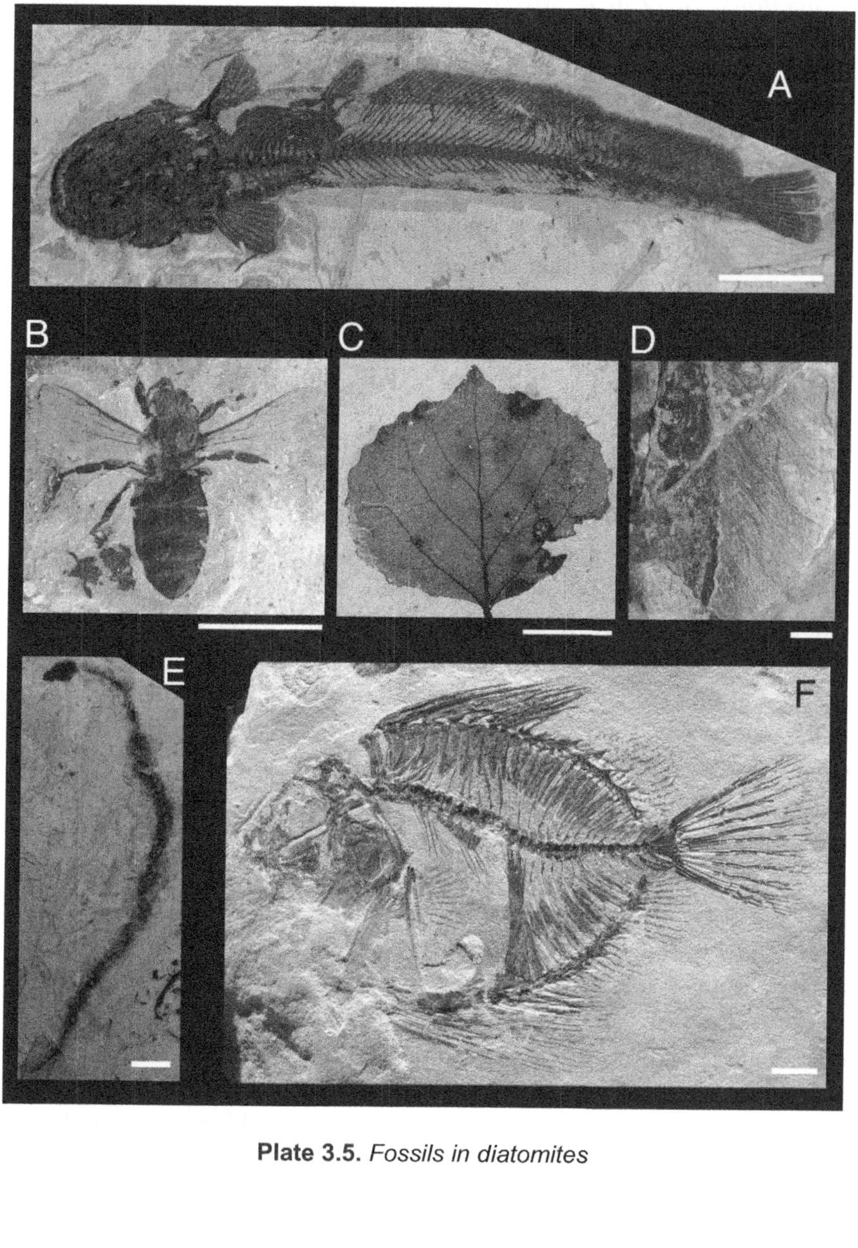

Plate 3.5. *Fossils in diatomites*

NOTE ON PLATE 3.5.– *A to E: Tortonian diatomites, Saint-Bauzile, Coiron, Ardèche (south-eastern France). A: Silurian fish; B: hymenoptera insect of the bee type; C: angiosperm leaf; D: hair of suidae; E: snake; F: Messinian diatomites, Oran, Algeria, John Dory (collections: A, D to F: MNHN, Paris; B and C: R. Jattiot; photos: D: L. Cazes; A, E and F: J.-P. Saint-Martin; B and C: R. Jattiot). Scale bars: A, D and E = 5 cm; B and F = 1 cm; C = 2 cm.*

3.2. Accumulations of teeth

The teeth of vertebrates, be they mammalian or reptilian forms, are much better fossilized than their bones. Teeth are dense, and already have a very high level of mineralization and crystallization during the animal's lifetime (fluorapatite $Ca_5(PO_4)_3F$ or hydroxyapatite $Ca_5(PO_4)_3(OH)$). They can be fossilized practically as they are, barring a few geochemical modifications of their component carbonates and phosphates. It is therefore not uncommon to find large numbers of fossilized teeth without the original jaws from which they came, the bones not having been preserved. In the majority of cases, these teeth have been transported, and brought together by currents (river or sea currents), in facies where the aquatic currents lose their energy: embankments, meanders and oxbow lakes on water courses, coastal sands, and protected zones in lagoons. The facies formed by these accumulations are either faluns, for fossils from the Cenozoic, or river delta or coastal microconglomerates, for cases earlier than the Cenozoic. Note that once they are fossilized, the color of the teeth may range from white to blackish brown, with gray, bluish, orangey or brown shades in between. These variations in color are attributable to a wide range of parameters, starting with the age of the teeth: those from the recent Quaternary tend to be white; those from the Cretaceous or the Jurassic are generally black; those from the Tertiary may be orangey or brown (though there are many exceptions to these rules of thumb). The burial time must be viewed in parallel to a chemical alteration involving impregnation by minerals from the surrounding sediment – for example, reddish, brown or dark brown coloration may indicate the presence of iron, black may indicate manganese, orange phosphate, etc. (this is by no means an exhaustive list) (Kohn et al. 1999; Schoeninger et al. 2003).

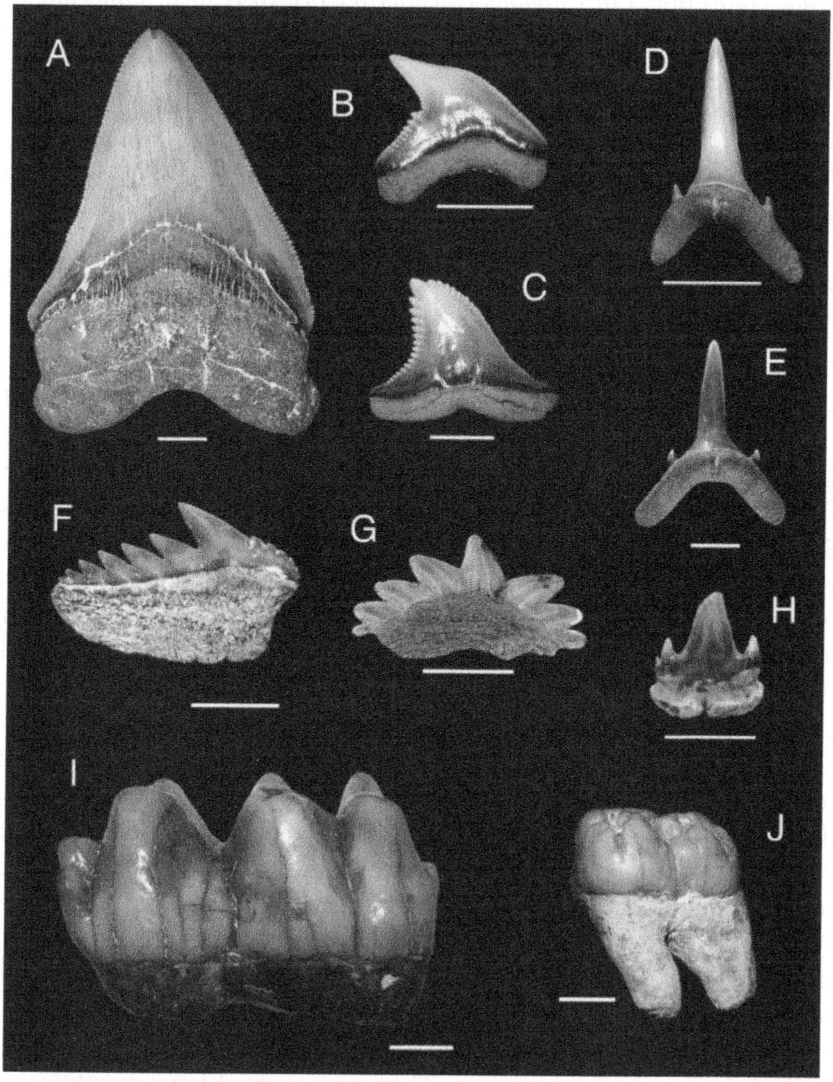

Plate 3.6. *Fossilized teeth from the Miocene faluns in Brittany (north-western France) – the fossilized teeth may vary in color, independently of the species from which they came, corresponding to shades of bluish gray or gray-beige (A to D), brown or orangey (E, I and J) or darker, with black parts (H).*

NOTE ON PLATE 3.6.– *The impregnation of the teeth with a chemical element (e.g. Fe or Mn) from the surrounding sediment is often the cause of this coloration. For example, consider the brown-reddish stains due to iron oxides (I). Sharks: A:* Otodus *(*Megaselachus*)* megalodon*; B:* Galeocerdo aduncus*; C:* Hemipristis serra*; D and E:* Carcharias cuspidata*; F and G:* Notorhynchus primigenius*; H:* Scyliorhinus distans, *proboscian mammal; I:* Gomphotherium angustidens, *sirenian mammal; J:* Metaxytherium medium *(collections: IGR; photos: J. Royer). Scale bars: A to D, F, I and J = 1 cm; E, G and H = 5 mm.*

Example of the Miocene faluns of Anjou, Brittany and Touraine (north-western France)

Faluns are non-consolidated, shelly, carbonated deposits, containing a proportion that varies in both quantity and particle size distribution of siliceous elements (sand, gravel, pebbles). For the most part, they were deposited in the shallow epicontinental seas which covered western France in the Miocene and Pliocene, collectively known as the “Faluns Sea”. These biodetritic facies were formed in deposit environments with some degree of agitation, leading to the concentration of shells (mollusks, brachiopods) and tests (sea urchins, foraminifera) of invertebrates, and the remains of teeth, and more rarely bones of coastal vertebrates. These remains come not only from the coastal marine environment (cartilaginous or bony fish, sirenians, cetaceans), but also from the land, with the bodies of carnivores, ruminants and primates having been carried to the coast by water courses. Within the faluns there are facies of lumachelles containing bivalves (ark clams, oysters) relatively poor in vertebrate remains, but they are the (micro)conglomeratic facies which are richest in fossilized teeth (see Plate 3.6). In addition to the Neogene deposits of western France, faluns deposits from the Cenomanian have been found in Charentes, south-western France (Vullo et al. 2003; Vullo and Néraudeau 2008). These deposits were generally formed in contexts of low sea level at the start of the marine transgression. They concentrated large quantities of debris of invertebrate marine organisms (various types of shells, spines and tests of sea urchins, carapaces of crustaceans, ossicles of asterids, tubes of annelids, etc.) and vertebrates (teeth, scales, coprolites, fragments of carapaces or eggshells), but also more land-dwelling species of plants (wood) and vertebrates (teeth, carapaces, bones). They are remarkable paleontological sources which could be described as "bio-accumulations".

3.3. Bone beds

The bones of mammals and reptiles are less well fossilized than their teeth. Bones are only really dense and properly mineralized on the edges, as the medullary part is vacuolar, with an interwoven mineral framework that is more or less perforated depending on the group in question. This fragility, and a relatively low rate of crystallization of the dense parts make these skeletal remains that are very vulnerable to post-mortem transport and acid attack from the soils. Thus, it is fairly rare to find facies rich in fossilized bones, and where it is the case, they are facies of very localized accumulation known as *bone beds*.

Example of the Berriasian bone beds from Charente (south-western France)

The boundary between the end of the Jurassic and the start of the Cretaceous (the Tithonian and Berriasian stages) was marked by a low sea level, as can be seen in numerous lagoon or near-coastal facies – notably in western France (Colin et al. 2004). In what are thought to have been lagoons and coastal marshes, periodically subject to flooding by the sea, waterways brought and accumulated the bodies of multiple reptiles, which the anoxic sludge buried and protected. The Angeac and Cherves sites in Charente (south-western France), are evidence of this process of build-up and fossilization of the bones of dinosaurs, turtles and crocodiles (see Plate 3.7A). In Cherves, within a gray, reducing clay matrix, we find brown/blackish fragments of bones and the osteoderms of crocodiles and turtles (see Plate 3.7A) (Buffetaut et al. 1989; Pouech et al. 2014). In Angeac, the accumulation is particularly spectacular, with thousands of bones overlapping over around 1 m of depth, across a surface area of at least several ares, containing specimens ranging from small phalanges measuring only a few centimeters to astounding femurs measuring over 2 m in length. In addition, the bone preservation at these sites is exceptional, with medullary parts not filled with sediment, either empty or partially epigenized, lending the bones an almost contemporary appearance, except for their brown-black color, at the base of the bone bed, due to the ligniteous and anoxic sediment surrounding them, which has preserved them, as is the case in Cherves. In the upper parts of the bone bed, which is more liable to alteration, brown/red tinges linked to iron oxides stain the surface and the interior of the bones (see Plate 3.7B) (Néraudeau et al. 2012; Allain et al. 2022).

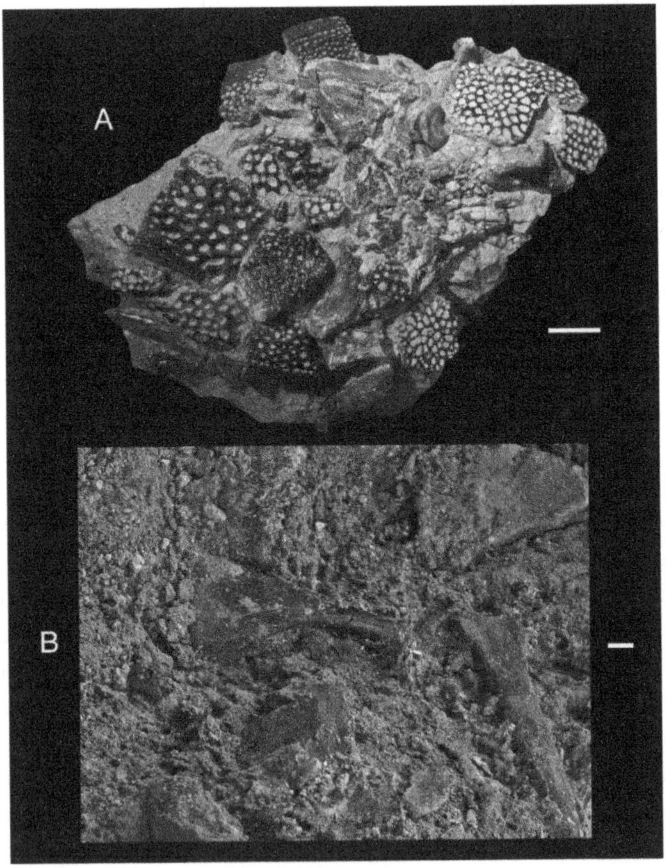

Plate 3.7. *Bone beds from the Berriasian of Charente, south-western France – the Purbeckian, lagunar or paralic facies from the Berriasian of Charente are rich in bone beds which contain the remains of crocodiles, dinosaurs and turtles*

NOTE ON PLATE 3.7.– *A: accumulation of remains of crocodiles and turtles, blackish in color, in the gray reducing clay, typical of the Cherves-de-Cognac site (south-western France); B: accumulation of dinosaur bones, along with fragments of turtles, in a clay microconglomerate with limestone gravel, typical of the Angeac site (south-western France); while the lower levels in Angeac contain the same blackened bones and osteoderms as those found in Cherves, the upper levels, which are more subject to alteration, show oxidation (of iron) that is beige, brown or reddish, on the vertebrate remains (collections: A: É. Dépré; B: in situ view; photos: D. Néraudeau). Scale bars: 1 cm.*

3.4. Fossilized tracks and trackways

In paleontology, the study of ancient biological traces is known as “paleoichnology”, or “palichnology” (from the Greek *iknos*, “trace”, and *logos*, “speech”). This discipline looks at all types of biological traces. It may be a burrow or a trace of locomotion left in the substrate by an invertebrate or a vertebrate. As the fossil of the organism itself is not typically preserved with the trace, it is often difficult to make a link between the two. In addition, a single animal can produce various types of tracks depending on its behavior and the nature of the sediment. Conversely, multiple species can leave the same trace when engaging in the same activity. In very rare cases, we may find the animal’s fossilized body at the end of a track of its movement (in this case, we refer to “Mortichnia” traces). Various examples of Mortichnia of horseshoe crabs, lobsters and fish have been found in the lithographic limestone of Bavaria (Germany) (Lomax and Racay 2012; Schweigert et al. 2016).

Traces of movement prove to be a very informative tool for paleontologists. On the basis of the shape and size of the trackway and of the tracks (tridactyl or pentadactyl, distance between and length of digits, etc.) (see Plates 3.8A and 3.8B), palichnologists can deduce a great deal of information about the nature of the animal, its dimensions, the way it moved, its behavior and the speed at which it moved. The depth and detailed relief of the tracks may also be analyzed using photogrammetry (see Box 3.1) to produce a reconstruction as detailed as possible of the anatomy of the feet or hands which made the tracks (number and length of phalanges, orientation of claws, etc.) (see Plate 3.8E to 3.8K).

When the substrate has been significantly disturbed by the passage of large tetrapods (notably in a very humid context), the outline of the footprints cannot be discerned, and we merely see a deformation (due to the piston effect) of the sedimentary layers "crushed" by the animals' feet. This deformation of deposits can be described as trampling and, in the specific case of dinosaurs, dinoturbation.

In order to be preserved, the tracks need to be covered quickly by sediment so as not to be altered by physical factors (water, wind, intense desiccation, etc.) or biological factors (e.g. bioturbation).

The morphology of the trace depends on a number of parameters, including the nature of the substrate and its level of water saturation, the weight of the animal, its shape and the position of its limbs. A clay-rich soil may yield a track with clear and detailed contours, whereas a print left in sand or gravel will be cruder. Although this general rule holds true, in certain particular sedimentary contexts, it happens that the footprints of vertebrates are remarkably well preserved in coarse sandstone or microconglomerates (Szewczyk et al. 2020), or breccia. In addition, the presence of protective microbial mats often plays a key role in the preservation of palichnological sites (Bernier et al. 1991; Seilacher et al. 2008). When the energy of the medium is low, a biological mat may develop in fine films of water (on a beach, a riverbank, a marsh) and on the surface of substrates containing footprints, thus protecting them from erosion until they are covered with sediment (Krumbein 1978; Freytet et al. 2003).

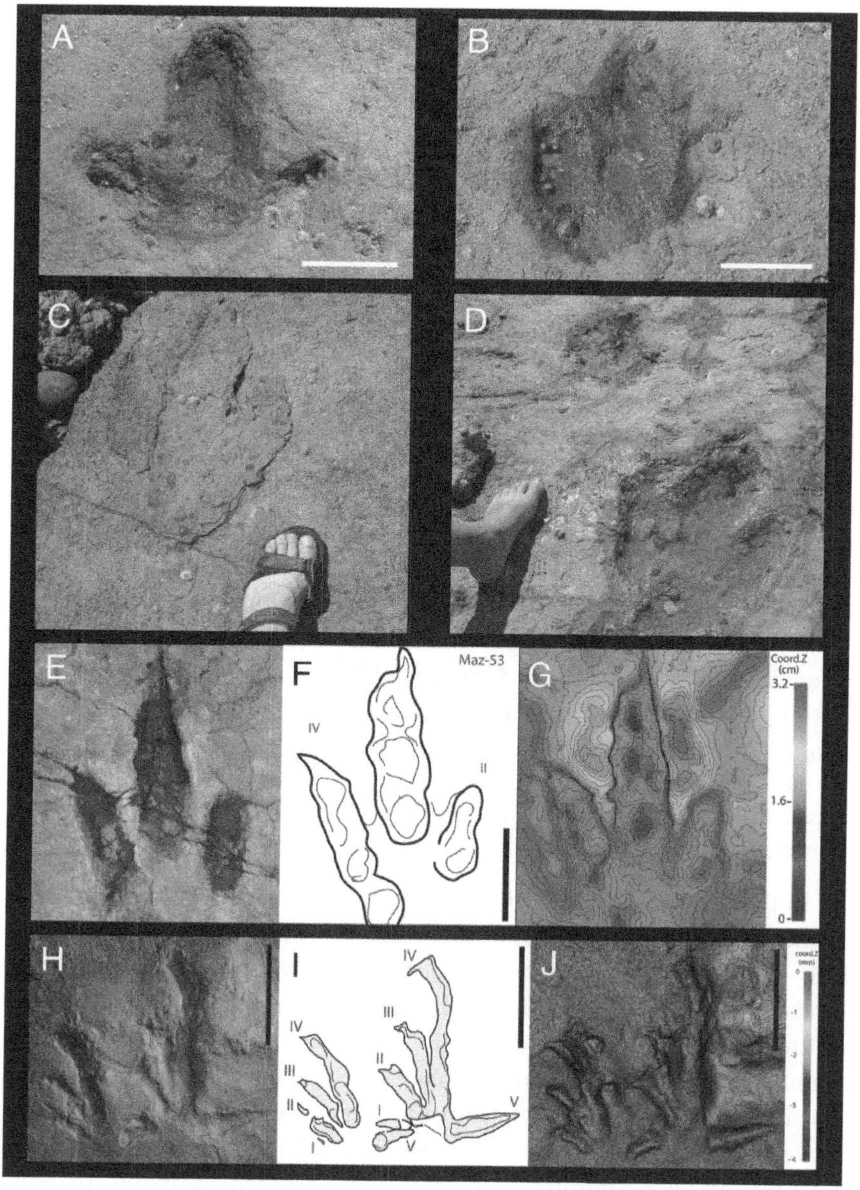

Plate 3.8. *Tracks of tetrapods*

NOTE ON PLATE 3.8.– *Damp substrates (lagunar, lacustrine, littoral, fluvial, marshy) have sometimes finely preserved the manus and pes tracks of the tetrapods which walked over them. Many examples can be found in France in the deposits from the Hettangian (Lower Jurassic), notably in Vendée, in western France (A to D) and Les Causses, in south-eastern France (E to G), but also Permo-Triassic deposits, such as the Permian deposits at Latour (Saint-Affrique region, Aveyron, south-eastern France). A and B: comparison between the respective footprints of two dinosaurs with tridactyl feet, one comprising a long central digit with two shorter lateral digits (A), the other characterized by digits that are more parallel and more similar in length (B), footprint of type* Eubrontes*; C and D: comparison between two* Eubrontes*-type tracks, one not demolded by erosion – i.e. still filled with the deposits from the layer above (C) – and the other released by erosion (D); E and I: tracks directly photographed; F and J: tracks recovered by drawing software; G and K: tracks imaged using photogrammetry, with negative reliefs being accentuated by the intensity of the blues, and positive reliefs being accentuated by the intensity of the yellows and oranges. I to K: pentadactyl track of* Dromopus *(collection: I to K: Musée de Lodève; photos: A to D: in situ, D. Néraudeau; E to K: J.-D. Moreau). Scale bars: 10 cm.*

Example of dinosaur footprints from the basal Jurassic in the Causses basin (south-eastern France)

In the south of the Massif Central, the Causses basin (south-eastern France) contains more than 60 tracksites with the tracks of archosaurs (the large majority traces of crocodylomorphs and theropod dinosaurs), dating from the Hettangian/Sinemurian (–201 to –191 million years) (Demathieu et al. 2002; Gand et al. 2007). This abundance makes it one of the places in the world with the highest densities of dinosaur tracks from the Early Jurassic. These traces of movement can testify to very varied behaviors, such as walking, running, or even swimming. Certain details, such as marks of paw pads or claws, are sometimes present, and the detailed reliefs can be demonstrated by photogrammetry (see Plates 3.8G to 3.8K). In much rarer cases, the markings of skin may be finely imprinted. For example, this type of exceptional preservation has been observed in theropod footprints from the Hettangian of Aveyron (south-eastern France). This exceptional state of preservation is largely due to the paleoenvironmental and sedimentary context of the track levels. These are essentially made of dolomite (a mixture of calcium- and magnesium carbonate) with algal laminites, formed in the tidal domain of a marine marsh, in which microbial and/or algal mats covered the substrate which was periodically emerged and then submerged. These protective mats provided a certain "consistency" to the sludgy substrate, favoring the very fine preservation of footprints.

In France, deposits from the Lower Jurassic, notably from the Hettangian stage, have produced multiple sites with fossilized reptile tracks. In addition to the Causses and other localities in the south-west of the country, there is an exceptional fossil deposit in Vendée (western France), on the Talmont-Saint-Hilaire coast (see Plates 3.8A to 3.8D).

Two other stratigraphic intervals – one at the Permian/Triassic transition (see Plates 3.8I to 3.8K), and the other at the Jurassic/ Cretaceous transition, contain tetrapod tracks (Gand 1993; Moreau et al. 2017c).

New imaging techniques, such as *photogrammetry*, have proved to be highly effective in paleontology (Mallison and Wings 2014). This is a technique using photographs which is able to create a 2D or 3D model of an object or surface with high definition. Photogrammetry can produce orthophotographs (scaled images with correction of distortion problems), useful for the detailed mapping of small objects (e.g. a fossil), and of very large surfaces (e.g. an outcrop). In addition, this technique is often able to reveal details that are invisible from the ground, and perform semi-automated measurements using the 3D models. Photogrammetry produces 3D "digital snapshots", and is therefore greatly beneficial from both a scientific and a geoheritage standpoint (see Plates 3.8G to 3.8K).

Box 3.1. *Photogrammetry – an imaging technique able to create a 3D model of a surface or object using photographs*

3.5. Fossils with anatomical connections

Certain organisms are fossilized almost in one piece, such as ammonites and gastropods, which have a continuous planispiral or trochospiral shape in their shell. Certainly, the internal casts of ammonites may break apart at the divisions between the chambers when the shell itself is not preserved, but where the shell is present, it holds the fossil together. On the other hand, other mineralized elements of these mollusks, such as the radula or beak (rhyncholite), are very rarely found alongside the fossilized shell, to which they are not strongly structurally attached.

In crustaceans, notably in decapods, dissociation is common, but the separate parts tend to be relatively large, usually with separation of the claws and the cephalothorax. However, the fossilization of the interconnected whole is relatively common in particular taphonomic contexts – notably when the animal's body is host to mineralizing bacterial activity which leads to the formation of a nodule (e.g. a phosphate nodule) which surrounds all or part of the carapace (see Plate 3.9A).

Conversely, in echinoderms, the skeleton is made up of a multitude of elements (articles, plates, ossicles, spines, mandibles), held together by putrefiable tissues whose post-mortem decomposition generally leads to the dislocation of the specimens which are later to be fossilized. In sea urchins, this leads to the separation of the test and spines (and sometimes the mandibles), and it is fairly rare to find a fossil in which they remain connected (see Plate 3.10D). In crinoids and certain Paleozoic echinoderms, the skeleton is a complex and fragile composition of a piece affixed to the substrate (the pseudoroot, known as the "base"), a peduncle in which dozens of articles are superimposed on one another, a composite calyx and arms which are also made up of a multitude of elements. The fossilization of all or a large part of this architecture in a single, almost complete fossil is absolutely exceptional (see Plate 3.10A and 3.10B). Finally, in the case of starfish, whose skeleton is a mosaic of tiny tile-like structures (ossicles) with very little interleaving, the fossilization of their complete, non-dissociated skeleton is highly exceptional (see Plate 3.10C).

Plate 3.9. *Various fossils with intact anatomical connections*

NOTE ON PLATE 3.9.– *A:* Potamon potamios *crab, mid-Pleistocene, Denizli, Türkiye, with the cephalothoracic carapace and pincers still connected; B: crocodilian* Macrospondylus bollensis, *Toarcian,*

Holzmaden, Germany, with the bony skeleton with its connections, and the preservation of some osteoderms on the dorsal part; C: a leafy branch of the conifer Geinitzia, *Cenomanian, Charente-Maritime (south-western France), showing cones at the end of the various branches (collections: A: J.-C. Le Sager; B: IGR; C: É. Dépré; photos: A: M. Guillo; B and C: D. Néraudeau). Scale bars: A and C = 1 cm; B = 10 cm.*

In the plant kingdom, the association, in the fossil state, of the different parts that made up an individual specimen is practically unknown. A complete tree extends from the roots to the very tips of the branches, including the trunk, and in certain seasons with leaf buds, fruits or cones, or indeed spores or pollen grains. However, these various elements are not fossilized in connection with one another, meaning that we do not see preservation of the natural contacts between, at least, roots-trunk-branches-leaves. In exceptional cases, we may indeed see connections between leaves and the extremities of the branches (leafy twigs), or with the associated reproductive structures – notably cones in the case of conifers (see Plate 3.9C).

For vertebrates, the different bones are usually fossilized "loosely", i.e. in isolation, in the wake of various forms of biological perturbation (e.g. carrion feeding) or physical perturbation (e.g. transport by water), where it is not just the teeth that have managed to withstand all biological, physical and chemical attacks. This is the case most commonly encountered with bone beds. However, when the majority of the skeleton is preserved complete with its anatomical connections at the site of its post-mortem burial, we see remarkable fossilization. In the case of fish, it is relatively commonplace in view of: firstly, their likelihood of being deposited in a horizontal position on the silty substrate; and, secondly, their relatively thin bodies, meaning they can be covered and buried by not too thick a layer of sediment (such is the case with laminar fish-bearing sediments: papery clays, diatomites, leafy limestones). For tetrapods, the three-dimensional layout of their bones, occupying a rather larger volume of space, means it takes longer for them to be completely buried, and it happens more rarely. Therefore, it is rather rare to find them preserved with their anatomical connections – all the more so with the maintenance of the osteoderms on the surface of crocodilian skeletons (see Plate 3.9B).

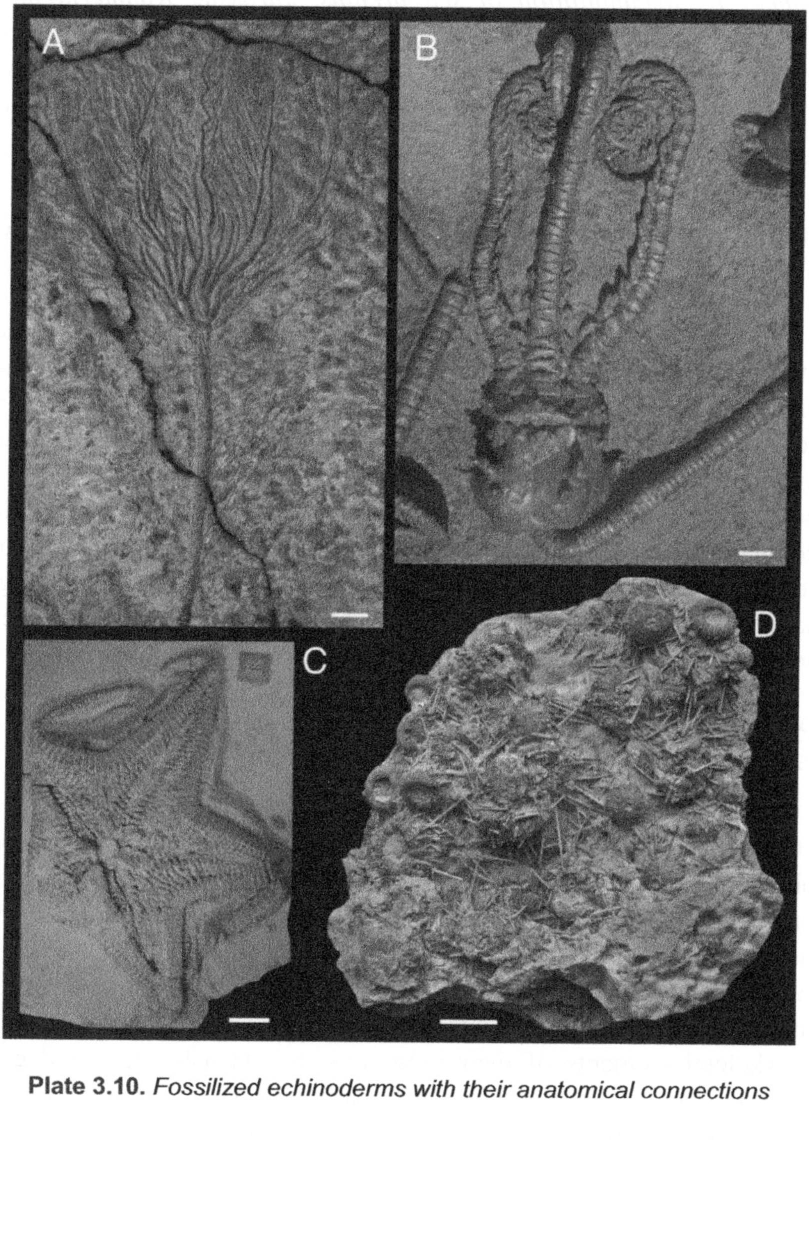

Plate 3.10. *Fossilized echinoderms with their anatomical connections*

NOTE ON PLATE 3.10.– *A: crinoid* Seirocrinus subangularis, *Toarcian, Holzmaden, Germany; B: echinoderm pelmatozoa (*Jimbocrinus bostocki*), Permian, western Australia; C: asterid (*Tethyaster guerangueri*), Cenomanian, Gazonfier, Le Mans, Sarthe, western France; D: accumulation of sea urchins, Jurassic, Kimmeridgian, Chassiron, Île d'Oléron, Charente-Maritime, south-western France (collections: A: IGR; B and D: É. Dépré; C: Musée Vert, Le Mans; D: Muséum d'Histoire naturelle de La Rochelle; photos: A and B: D. Néraudeau; C: N. Morel; D: R. Vullo). Scale bars: 1 cm.*

3.6. Eggs and trophic traces (bite marks, galls, regurgitations and coprolites)

A fossil may be remarkable, or even exceptional, in terms of the fine preservation of its anatomy, its most fragile tissues or (and) the originality of its mineralization. However, certain fossils, other than their intrinsic quality, may be exceptional in terms of the part of the trophic network, or biological activity, to which they attest. In some cases, they document moments in time, snapshotted in the sedimentary rock, of interactions among species (feeding, predation, parasitism) or among individuals of the same species (brooding, mating, birthing). In others, they extend the fossil record of a species by preservation of its excrement, its eggs or its regurgitations.

Example of vertebrate/invertebrate interactions or interactions between invertebrates

Direct food-chain interactions are mainly documented by traces of bites on the shells or bones, or traces of grazing on leaves or wood (phytophagy, xylophagy). Thus, we find the teeth of marine reptiles or sharks embedded in the shells of ammonites that they have attacked or eaten (see Plate 3.11G). In addition, regular sea urchins have jaws (Aristotle's lantern) which they use to scrape off algal films or other fine organic tissues. The five teeth in these very powerful jaws scrape the skeletal elements of their prey (oysters, starfish, etc.), and can easily be recognized by their starlike structure which reflects the sea urchin's own pentaradial symmetry, corresponding to *Gnathichnus-*

type traces (see Plate 3.11F) (Breton et al. 2017). Feeding interactions may also be documented directly by the fossilization of a meal an animal has eaten, still in place in its digestive system. For example, in the abdomen of a marine reptile such as an ichthyosaur, we can identify the rostra of belemnites it had eaten, "bathed" in the black ink produced by these cephalopods (see Plate 3.12).

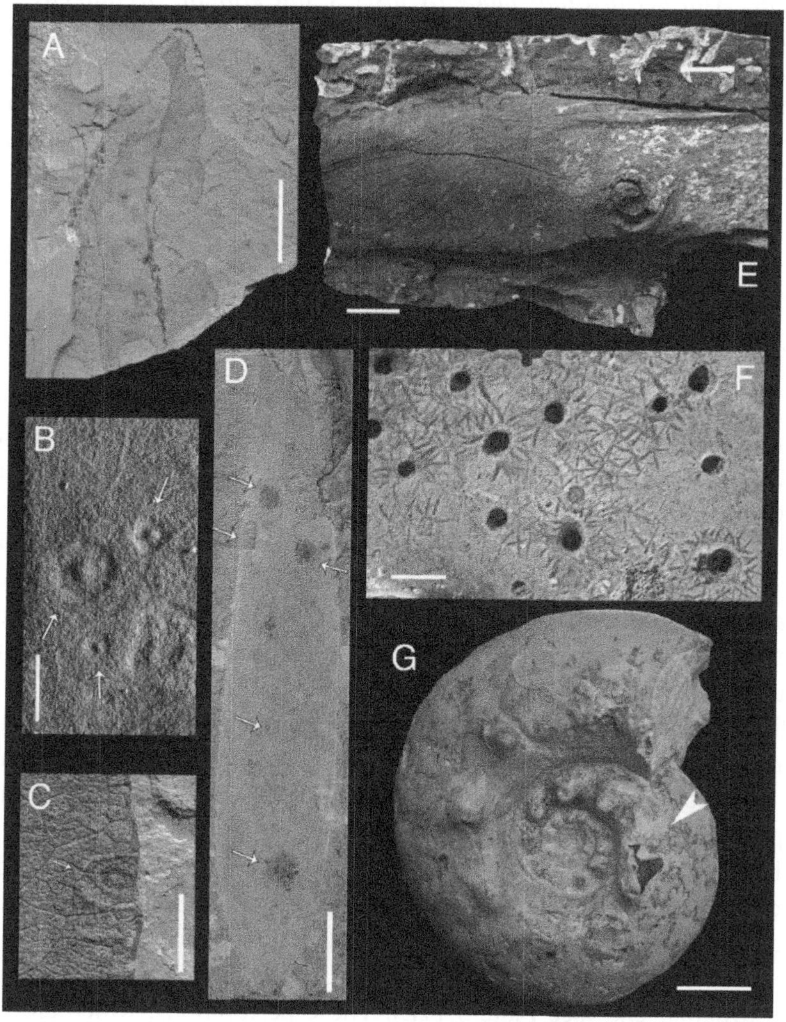

Plate 3.11. *Various trophic interactions*

NOTE ON PLATE 3.11.– *A to D: leaves from the Cenomanian of Tonnay-Charente, Charente-Maritime, south-western France. A: angiosperm leaf with indentations at its edge, indicating folivorous activity; B and C: fragments of the limb of an angiosperm leaf bearing galls (indicated by the arrows); D: leaf from the conifer* Dammarophyllum *with circular marks from perforation and suction; E: fossilized wood in the form of lignite, with perforations made by insects, late Albian, Archingeay-Les Nouillers, Charente-Maritime, south-western France; F: star-shaped traces of predation made by the jaw of a sea urchin on an oyster shell, Cenomanian, Sarthe, western France; G: ammonite* Orthaspidoceras *with a shark tooth encrusted in its shell, following a bite, Upper Kimmeridgian, Yves, Charente-Maritime (collections: A to E: IGR; F: Musée Vert, Le Mans; G: Muséum de La Rochelle; photos: A, E: D. Néraudeau; B to D: T. Wappler; F: G. Breton; G: R. Vullo). Scale bars: A, C to E = 1 cm; B = 5 mm; F =2 mm; G = 2 cm.*

Example of interactions between plants and insects

In laminar clays and limestone, or lithographic limestone, the leaves of angiosperms, conifers and ferns can be fossilized in particularly fine detail. This type of preservation is favorable to the preservation of all traces of "attack" undergone by the plants "during their lifetime", or after the leaves have been detached from the branch – notably by insects. Firstly, this is the case with traces of *folivory*: very often, leaves bear notches on their edges, or perforations of the limb, which indicate they have been consumed by grazing or suction by arthropods, either adult or larval (see Plate 3.11A to 3.11D). Just like the traces of folivory, leaves fossilized in fine sediments may be preserved with the marks of *parasitism*. These may include circular galls or tunnels made in the thickness of the limb (see Plate 3.11B and 3.11C). The galls appear as disks or ovals which are darker than the plant remains, with bulges, while the tunnels, which are also darker in color, are straight, circular or sinuous (Santos et al. 2022). Finally, both in the fossil record and in today's world, wood may show the hallmarks of insect attack, in the form of holes on the surface or centripetal tunnels with a more or less irregular end (see Plate 3.11E). These perforations, which are due to land-based insects, must not be confused with those made by teredos in the aquatic environment. The latter manifest on the surface of fossilized wood as holes that are similar to those made by insects, but the tunnels tend to increase in diameter inside the wood, and end in a "bulb" or cavern, at the site occupied by valves of the burrowing bivalve (of the genus *Teredo*) (Robin et al. 2018).

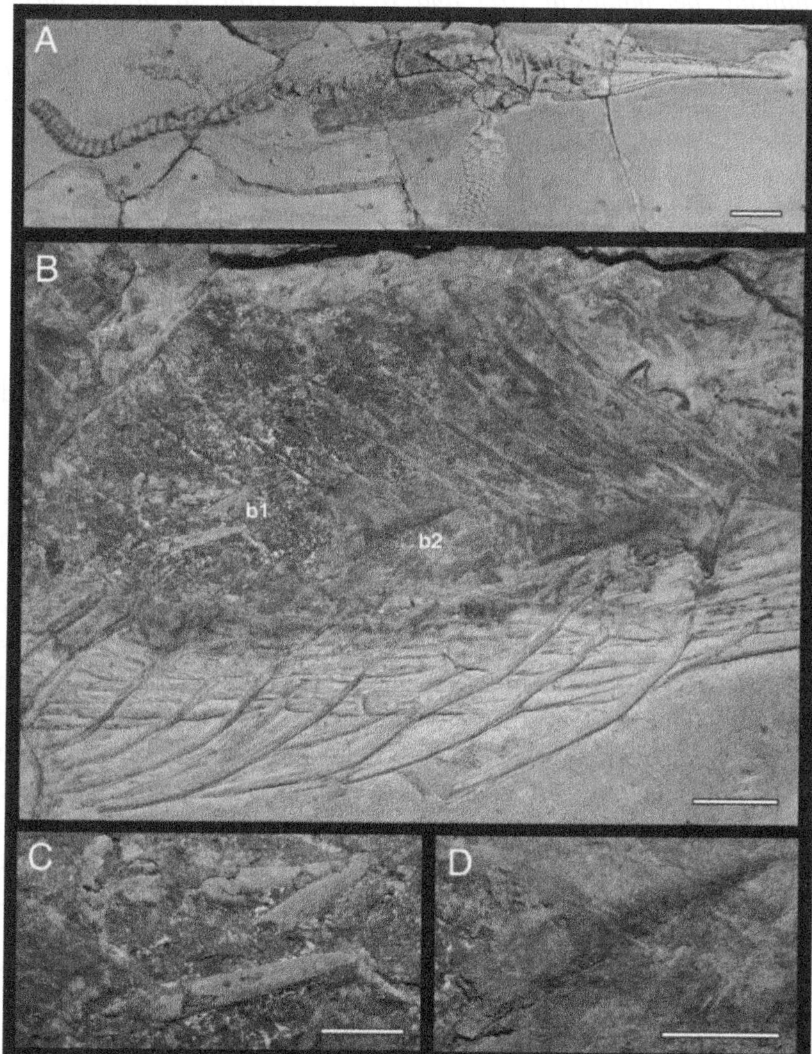

Plate 3.12. *Fossilized ichthyosaur still containing its last meal of belemnites*

NOTE ON PLATE 3.12.– *Ichthyosaur skeleton with anatomical connections, containing belemnites in the region of the stomach. Lower Jurassic, Holzmaden, Germany. A: overview of the skeleton, with the stomach having been blackened by the ink produced by the belemnites it has ingested; B: close-up of the stomach region, containing belemnites at (b1) and (b2); C: close-up of the three belemnites located at (b1) in image (B); D: close-up of the belemnite located at (b2) in image (B), with the phragmocone area visible in brownish-red, on the left-hand side (collection: IGR; photos: D. Néraudeau). Scale bars: A = 10 cm; B = 3 cm; C and D = 1 cm.*

Fossilized elements which indirectly document the diets of the species can also be found in fossilized excrement, known as "coprolites", or fossilized regurgitations, known as "regurgitalites".

Example of coprolites

The fossilized excrement of land-dwelling vertebrates can be found relatively frequently – notably that of hyenas – on the sedimentary floors of Pleistocene caves. For aquatic or semi-aquatic vertebrates, such as sharks and turtles, their coprolites are found in accumulations of microremains of vertebrates, such as the Cretaceous deposits rich in selachian teeth and plates from the carapaces of turtles (see Plate 3.13E and 3.13F). Aquatic or semi-aquatic invertebrates, such as crustaceans – notably decapods – also produced large quantities of small coprolites known as *Tomaculopsis*, fossilized in microconglomerates with teeth and small bones, and gathered together in cylindrical phosphatized groups, or in pancake-like form (see Plate 3.13C and 3.13D) (Breton 2011). Other invertebrates – notably certain annelids – produced traces, known as *Lepidenteron*, rich in debris from fish (scales, small bones) that they consumed, forming structures that fall somewhere between burrows and coprolites, present in the chalk or flints from the Upper Cretaceous (see Plate 3.13A) (Bieńkowska-Wasiluk et al. 2015). In land-dwelling invertebrates, fossilized excrement is particularly well documented in that of termites, preserved in ligniteous deposits, in amber or inside wood, dating from the Mesozoic or Cenozoic. Termite coprolites can be easily recognized by their regular shape with a hexagonal cross-section (Colin et al. 2011; Moreau et al. 2019b) (see Plate 3.13B).

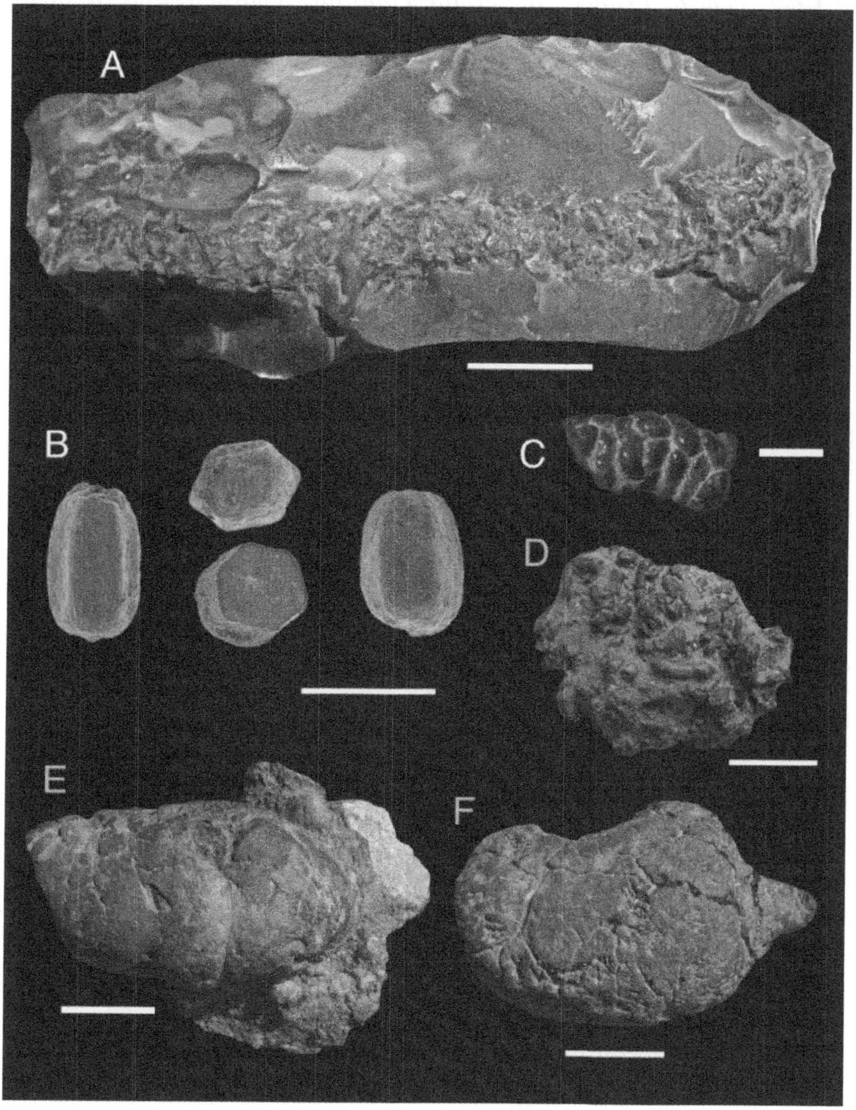

Plate 3.13. *Various coprolites from both vertebrates and invertebrates*

NOTE ON PLATE 3.13.– *A: coprolite of the type* Lepidenteron, *containing numerous fish scales, generated by large annelids, Turonian of Cher, central France; B: termite coprolites characterized by their regular geometric form and hexagonal cross-section, Cenomanian of Charentes, south-western France, SEM view; C and D: coprolites of the type* Tumaculopsis, *cylindrical (C) or pancake-shaped (D), made by decapod crustaceans, Cenomanian of Tonnay-Charente, Charente-Maritime, south-western France; E: shark coprolites, in spiral rings, Cenomanian of Fouras, Charente-Maritime, south-western France; F: pedunculate coprolite, attributed to an unknown reptile, Cenomanian of Fouras, Charente-Maritime (collections: A: K. Chartier; B to F: Néraudeau-IGR; photos: A, C and D: D. Néraudeau; B: J.-P. Colin). Scale bars: A = 2 cm; B = 1 mm; C = 2 mm; D to F = 1 cm.*

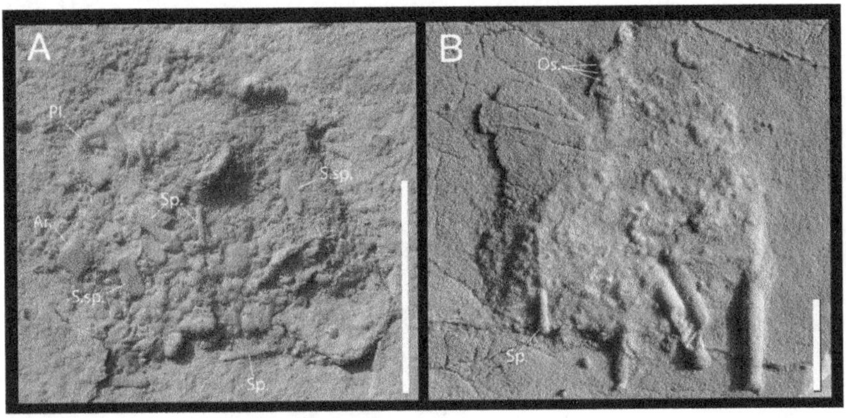

Plate 3.14. *Regurgitalites*

NOTE ON PLATE 3.14.– *Fossils of regurgitations, Kimmeridgian of Nivoliers (Causses, Lozère, south-eastern France), produced by malacophage fish which had eaten sea urchins, brittle stars and starfish, and regurgitated their skeletal elements: Ar = fragment of the Aristotle's lantern of a regular sea urchin; Sp and S.sp = different types of spines of regular sea urchins; Os = ossicles of starfish;*

Pl = plate of a sea urchin (collection: APHPL; photos: J.-D. Moreau). Scale bars: 1 cm.

Example of regurgitalites

Strange as it may seem, the regurgitations of certain animals can be fossilized reasonably well, and produce fossils that are exceptional in terms of their contents and what those contents actually signify.

The best known, both in the modern world and in recent paleontological contexts (Quaternary, Pliocene), are the regurgitations of birds of prey in the form of pellets, particularly abundant in prehistoric caves, indicative of small mammals (containing the teeth of rodents or insectivores), notably from the Pleistocene.

Regurgitations of malacophage fish which consumed either mollusks or echinoderms are well known in the lithographic limestone of the Upper Jurassic (see Plate 3.14) (Moreau et al. 2022) or the chalky deposits of the Upper Cretaceous (Breton 1992).

Rarer are fossils of regurgitations of Paleozoic tetrapodomorpha (Robin et al. 2022), and there are even surprising cases of regurgitalites of nautiloids (Silurian, –444 to –419 Ma) from Poland (Brachaniec et al. 2016).

Example of eggs

Eggs laid by vertebrates, and eggs laid by invertebrates, may be fossilized in exceptional deposits. In the case of the eggs of terrestrial vertebrates, rapid burial is crucial in order for the clutch to be protected from predators, from trampling, and from physical and chemical degradation due to exposure to wind, rain and UV, or to climatic temperature variations.

However, the original mineralization of the eggshells, which combine carbonate, magnesium and phosphorus, means they have a fairly good chance of being fossilized. We may find fossils of isolated eggs, or of whole nests of eggs, for different types of vertebrates, for both avian and non-avian dinosaurs (see Plates 3.15A and 3.15B) (Garcia 2000; Garcia and Vianey-Liaud 2001), or other reptiles such as turtles (see Plate 3.15C).

In aquatic vertebrates, eggs are less commonly documented. Usually, we find strings of eggs laid by fish and preserved in fine-grained laminar sediments, such as lithographic limestone (see Plate 3.15D) (Moreau et al. 2022).

More rarely, we may find fossils of shark egg capsules, notably in layered clays. Despite their softness and lack of mineralization, these structures have been preserved in the reducing silt which rapidly buried and impregnated them, probably with bacterial action (see Plate 3.15E) (Vullo et al. 2013).

3.7. Land-based palynomorphs: spores and pollen grains

Not all fossilized organisms are visible to the naked eye. Consider the case of microfossils, which measure only a few tens or a few hundreds of microns.

Of these, those which we call terrestrial palynomorphs are the most abundant forms in the deposits formed in terrestrial, freshwater or coastal marine environments. Indeed, there may be thousands of palynomorphs present in a few cubic centimeters of sediment.

In order to be sampled and identified, they first need to be extracted from the rock by chemical erosion (see Box 3.2). They are fairly commonly found preserved due to the highly resistant molecules that make up their walls – notably sporopollenin.

Just like teeth in vertebrates, palynomorphs are the remains which are most easily and most durably fossilized from plants. They include pollen grains and spores.

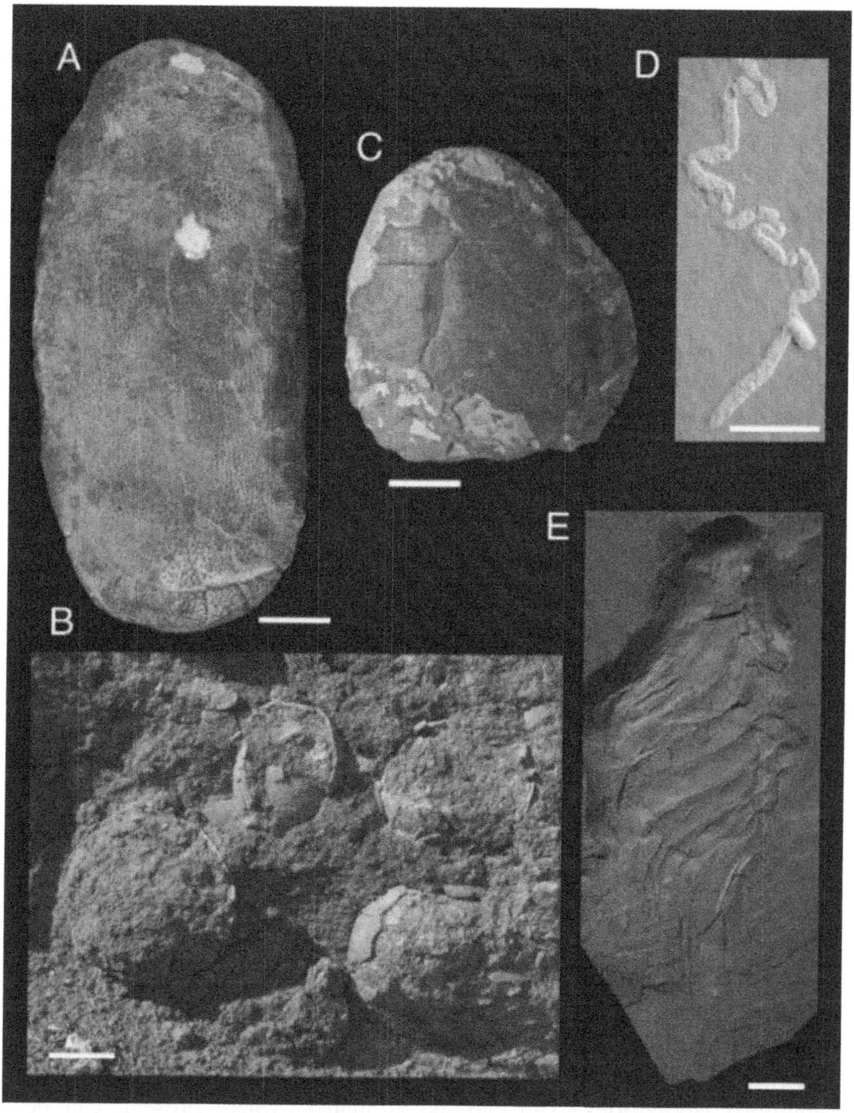

Plate 3.15. *Various eggs and clutches of eggs*

NOTE ON PLATE 3.15.– *A:* Oviraptor *egg, Cretaceous, Mongolia; B: nest containing several dinosaur eggs, Cretaceous, Kyrgyzstan; C: turtle egg, Aquitanian of Cantal, central France; D: bead-like clutch of eggs attributed to a fish, Kimmeridgian of Nivoliers (Causse Méjean, Lozère, south-eastern France); E: shark egg capsule, Cenomanian of Tonnay-Charente, Charente-Maritime, south-western France (collections: A: É. Dépré; B: in situ; C: IGR; D: APHPL; E: Muséum de La Rochelle; photos: A and C: D. Néraudeau; B: M. Jolivet; D: J.-D. Moreau; E: R. Vullo). Scale bars: A to C = 2 cm; D = 1 cm; E = 1 cm.*

Grains of pollen, produced both by conifers since the late Paleozoic and by angiosperms since the Cretaceous, are palynomorphs measuring between a few microns to a tenth or at most a fifth of a millimeter (see Plate 3.16). Spores, which are typical of ferns in the broad sense (see Plate 3.17), have been produced since the first plants colonized the land in the mid-Paleozoic. The wall of these terrestrial palynomorphs is made up of different layers, but it is the exine membrane, rich in sporopollenin, which is fossilized as the cellulosic layers degrade more easily (Bignot 2001). Sporopollenin is extremely resistant to all types of physical and chemical attack, be it by acids or bases, solvents, UV or desiccation. It is a biopolymer formed of aromatic carboxylic acids (aromatic carbon ring molecules) and aliphatic components (carbon chains, either linear or cyclic, but not aromatic).

Palynomorphs are generally fossilized in organic form in sedimentary rocks containing clay, marl or shale. Their *chemical extraction* involves dissolving and thereby removing the mineral matrix in which they are contained using very strong acids capable of destroying the gangue, without damaging the exine – the highly resistant organic envelope of palynomorphs. Hydrofluoric acid, which is one of the most corrosive agents, is the most widely used to destroy silica, but this is a dangerous practice and must only be carried out by trained staff in properly equipped laboratories. The carbonated matrix, where present, can be dissolved by somewhat less potent solvents, such as hydrochloric, nitric or sulfuric acid. Concentrated oxygenated water can also be used for recent deposits, to break up or eliminate the fragile organic material (foliar tissues, wood) accompanying the palynomorphs.

Box 3.2. *Techniques for chemical extraction of palynomorphs*

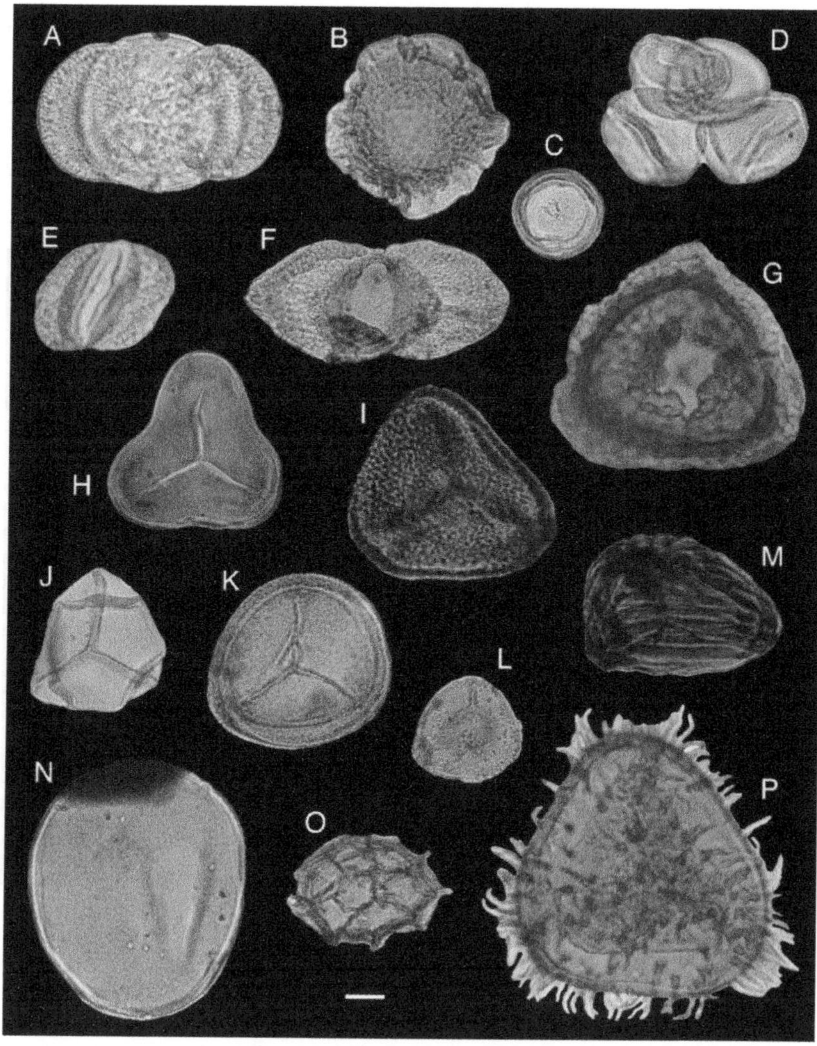

Plate 3.16. *Cretaceous pollen grains. Palynomorphs collected in the Berriasian deposits of Angeac, Charente, SW France*

NOTE ON PLATE 3.16.– *A:* Abietineaepollenites *sp.; B:* Callialasporites *sp.; C and D:* Classopollis torosus*; E:* Vitreisporites pallidus*; F:* Podocarpidites ellipticus*; G:* Aequitriradites verrucosus*; H:* Cyathidites australis*; I:* Microreticulatisporites diatretus*; J:* Gleicheniidites apilobatus; *K:* Densoisporites microrugulatus; *L:* Foraminisporis wonthaggiensis; *M:* Cicatricosisporites hallei; *N:* Ovoidites spriggi; *O:* Retitriletes *sp.; P:* Pilosisporites trichopapillosus *(collections and photos: D. Batten and F. Polette, University of Aberystwyth; photographic arrangement altered from Polette et al. 2018). Scale bars: 10 μm.*

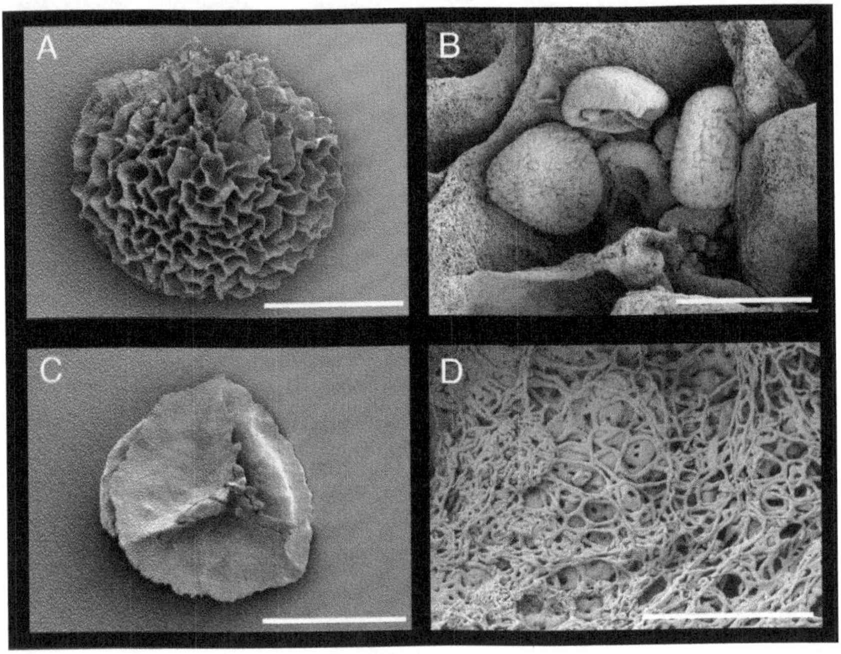

Plate 3.17. *Cretaceous spores collected in the Berriasian deposits of Angeac, Charente, south-western France*

NOTE ON PLATE 3.17.– *A: megaspore* Erlansonisporites *sp.; B: close-up of the reticular structure showing microspores lodged in the reticulum; C:* Minerisporites *sp.; D: megaspore* Erlansonisporites *sp.: high-magnification detail of the reticular structure of the external wall, composed of sporopollenin (collections and photos: D. Batten, University of Aberystwyth; photographic arrangement altered from Batten et al. 2010). Scale bars: A and C = 200 μm; B = 20 μm; D = 5 μm.*

3.8. Aquatic palynomorphs

The term "palynomorph" is not limited to plants' reproductive micro-components, such as spores and pollen grains. It is also used for aquatic microfossils, such as acritarches and chitinozoans from the Proterozoic and Paleozoic (Munnecke and Servais 1996; Paris 2006), and the cysts of dinoflagellate algae (dinocysts), present in various Mesozoic and Cenozoic deposits (Fauconnier and Masure 2004; Polette 2019). These microfossils, with organic walls, are highly resistant, and may be present in large quantities within a small volume of rock.

Chitinozoans are organic microfossils, whose biological origin (eggs of soft-bodied invertebrates?) and systematic position (arthropods? mollusks? other?) are still subject to debate. However, their usefulness is perfectly well recognized for dating marine sediment layers lain down between the start of the Ordovician and the end of the Devonian (Paris et al. 1996, 2004). There may be hundreds, or even tens of thousands of them in a single gram of rock. Their form is bladder-like, in the shape of an open-ended bottle, isolated, or clustered together (Paris and Nõlvak 1999) (see Plate 3.18). The strength (and elasticity) of their organic wall has allowed for their preservation in clays and limestones, phosphates and siliceous or ferruginous rocks. Their molecular composition has not yet been perfectly determined, but contrary to what the name indicates, fossilized palynomorphs have not preserved perfectly identifiable chitin (Paris 2006; Jacob et al. 2007). However, the constitutive elements attest to thermal changes undergone by the sediment surrounding them: when their geological context was raised to high temperatures, the result is the darkening and fragilization of their walls. These changes correspond to the maturation of their constitutive organic material and an increase in the associated carbon content. Their species are differentiated on the basis of the type, amplitude and position of their ornamentation (smooth, granular, tubercular, spiny, tentacular) (see Plates 3.18A to 3.18F). Chitinozoans are often found in isolation, but have also been known to form planar or tubular helical arrangements (see Plates 3.18G and 3.18H) (Paris and Nõlvak 1999).

Acritarches emerged in the Precambrian, around 1.8 billion years ago. In all likelihood, they are not a homogeneous group, as some forms are considered to be cysts of eucaryotic algae, comparable to dinoflagellate cysts (see Plates 3.19A and 3.19B), while others can be interpreted as being single-celled organisms (protists). The preservation of these extremely ancient aquatic palynomorphs is dependent on their sedimentary surroundings, and the diagenetic processes which they have undergone. Mennecke and Servais (1996) demonstrated 3D preservation of acritarches in fine carbonated matrices, which differ from the more "compacted" types of fossilizations which we observe in clays and shales, where the microfossils are more or less crushed. The same observation can be made for chitinozoa, which are compacted or even crushed when fossilized in clays, but preserved with full relief in siliceous rocks (cherts), limestone, phosphates or ferruginous rocks (Paris 2006).

Dinoflagellates are microscopic single-celled algae which have been around for over 400 million years. Ranging from a few tens to a few hundred microns in size, they have a "theca", formed of plates of cellulose. However, in the fossil state, their theca, which is very fragile, is generally not preserved, and it is their encystment with an organic wall, inside of the theca, which is fossilized (Bignot 2001) (see Plates 3.19C to 3.19G). From a chemical point of view, present-day organic dinocysts are composed of a resistant biopolymer, dinosporin, which, like the sporopollenin of terrestrial palynomorphs, plays a critical role in their preservation in ancient sediments (Kokinos et al. 1998). This potential for preservation is simultaneously linked to biomacromolecules with multiple carbon–carbon bonds, and the presence of aromatic molecules, which are very resistant to putrefaction. It should be noted that such long molecules with aromatic groups have also been seen in the composition of the walls of acritarches (Arouri et al. 2000) and chitinozoans (Jacob et al. 2007). Long carbon-chain polymers, notably those including aromatic ring groups, are now known as highly resistant molecules, the accumulation of which is harmful to the environment (pollution) and health (Bidaud 1998). In the context of fossilization, they have served as an asset for the preservation of the smallest and most ancient palynomorphs.

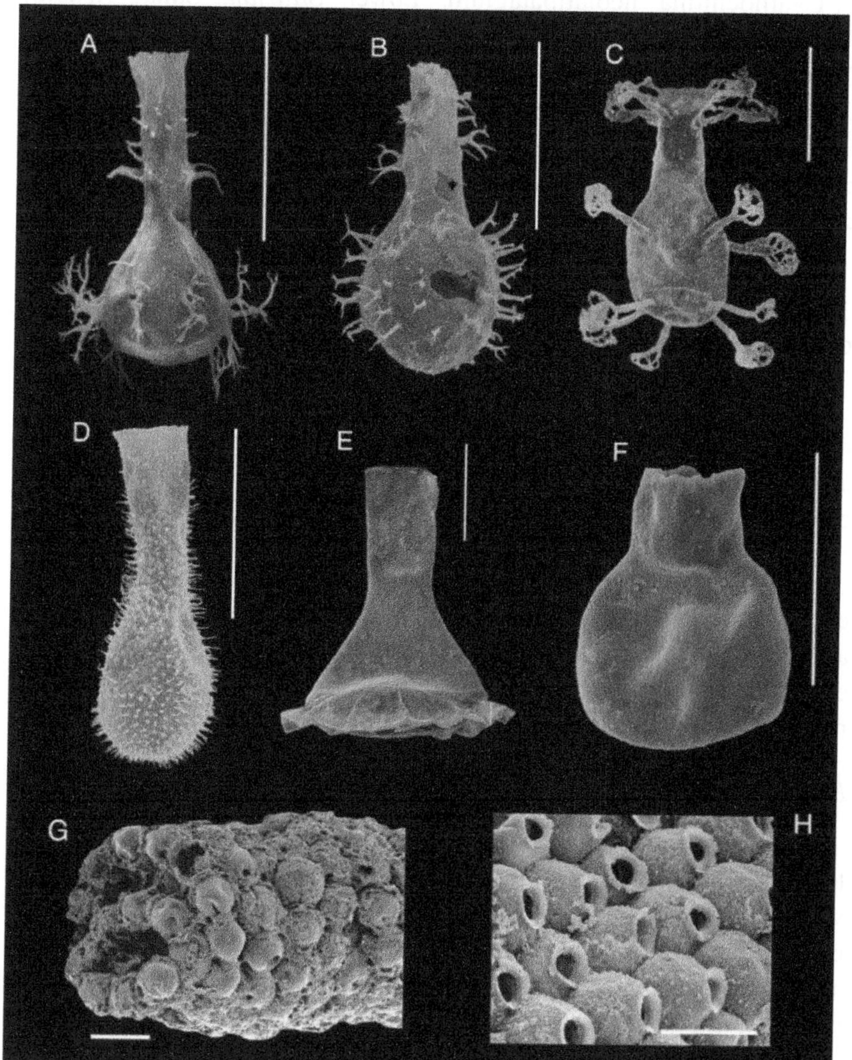

Plate 3.18. *Chitinozoan palynomorphs, extracted from Paleozoic rocks*

NOTE ON PLATE 3.18.– *A to F: Lagenochitinidae. A:* Ancyrochitina desmea, *Upper Silurian, Gotland, Sweden; B:* Ramochitina *sp., lower Devonian, Algeria; C:* Alpenachitina petrovinensis, *mid-Devonian, Brazil; D:* Angochitina longicoli, *mid-Silurian, Gotland, Sweden; E:* Cyathochitina neolatipatagium, *Lower Silurian, Saudi Arabia; F:* Lagenochitina prussica, *Upper Ordovician, Libya. G and H: Desmochitinidae. G: group (catenary helical structure) of* Desmochitina nodosa*; H: aggregation of* Desmochitina nodosa *interpreted as being a clutch of eggs (A to F) (collection: IGR; photos: D. Bernard and F. Paris; G and H: reproduced from Paris and Nõlvak 1999). Scale bars: 100 μm.*

3.9. Aquatic microfossils: algae, foraminifera and ostracods

Though tiny, a number of aquatic microfossils can be seen with the naked eye, or at least some of their species, which may reach sizes of around a millimeter. Such is the case, notably, for foraminifera and ostracods.

Other aquatic microfossils are too small to be seen with the naked eye, but may be present in such enormous numbers that, by accumulating, they actually constitute particular types of rocks. For example, diatoms produce diatomites; coccoliths make up chalk; and radiolaria are abundantly present in radiolarites (Bignot 1982, 2001).

Example of diatoms

Diatoms are microscopic brown algae, measuring between a few tens and a few hundred microns, usually round or elliptical, sometimes with a more complex and less rounded form, but almost always exhibiting bilateral symmetry (see Plates 3.20C to 3.20H). They have a siliceous skeleton, the frustule, which is the structure that is generally preserved when the organisms are fossilized. This siliceous framework is initially hydrated and included in an organic matrix, but only the silica in the diatoms can properly withstand the vicissitudes of fossilization (Bignot 2001). This amorphous silica is a sort of "biological glass", which is highly resistant to oxidation and bacterial action. The accumulation of frustules produces a siliceous sediment known as diatomite (see section 3.1.4).

Stratified, or laminar, diatomites themselves are rocks favorable to the preservation of other organisms which become trapped between their sedimentary layers, such as fish, insects and plants.

Diatomites, and consequently diatoms, are particularly abundant and well preserved in the Miocene deposits of the Mediterranean Basin, which attest to major variations in salinity, such as the Messinian Salinity Crisis (Pestrea and Saint-Martin 2002) (see Plates 3.20C to 3.20H).

Exceptionally, diatoms are part of aquatic micro-organisms which can be transported by sea spray and stick to flows of resin from conifers growing on the coast. Stuck within these flows, they are preserved, and ultimately become microfossils in the resin, which has become amber (see Plate 3.20A).

However, care must be taken when identifying them as microfossils in amber, as present-day diatoms may have encrusted amber nodules collected from damp sediments or have been washed (see Plate 3.20B). Thus, only those which have been found within the hardened resinous material are indisputably fossils.

Diatoms may also be fossilized by pyritization, in three different ways: their skeleton, made of opal-A, may be epigenized into pyrite; pyrite crystals may develop in the internal cavity of the frustules; or, after the dissolution of the diatoms in a hardened sediment, the void left by their dissolution may be filled with pyrite, forming a replica of the original organism. These processes have been observed in the Paleogene deposit of Saint-Josse, in Pas-de-Calais, northern France (Van Eetvelde and Dupuis 2005).

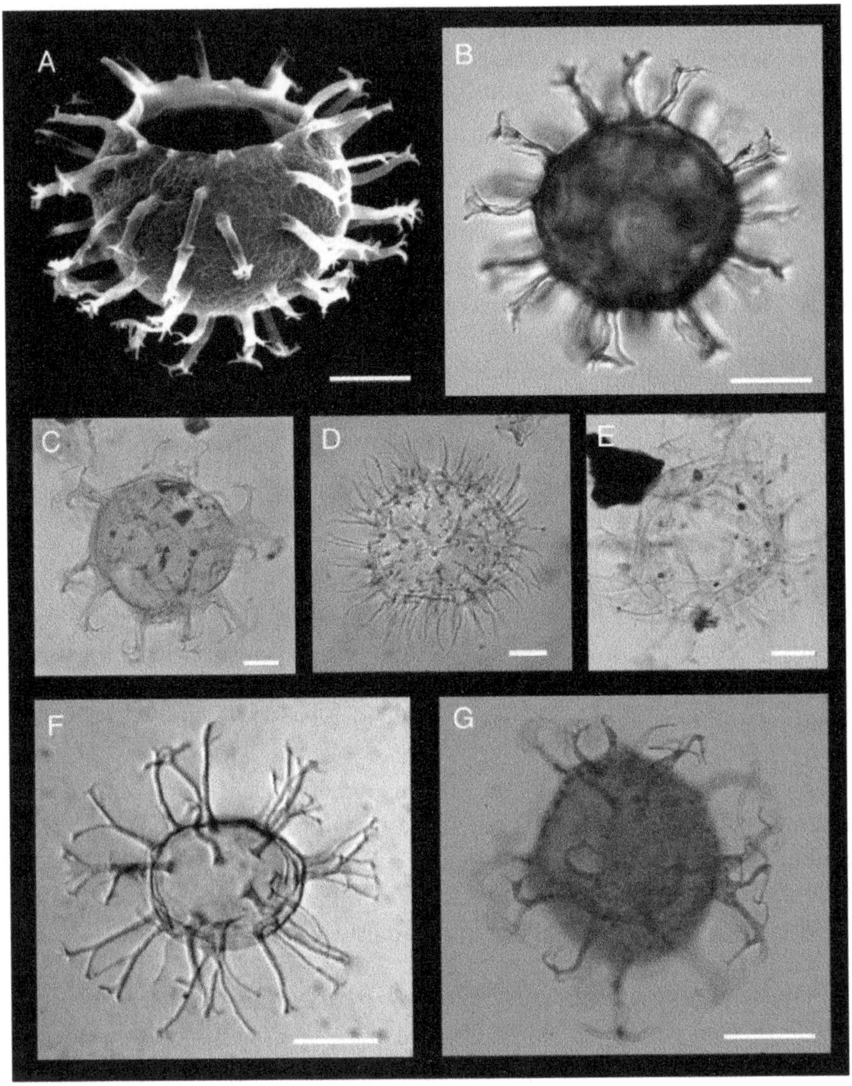

Plate 3.19. *Aquatic palynomorphs: acritarches and dinoflagellates*

NOTE ON PLATE 3.19.– *A and B: acritarches. A: SEM view of an acritarch* Stelliferidium, *Ordovician of Algeria; B: optical microscope view of an acritarch* Peteinosphaeridium, *mid-Ordovician of Öland,*

Sweden. C to F: dinoflagellate cysts (dinocysts), from the mid-Cretaceous. C to E: Cenomanian, Charentes, south-western France; F: Albian, optical microscope; C: Achomosphaera ramulifera*; D:* Coronifera oceanica*; E:* Spiniferites ramosus*; F:* Surculosphaeridium longifurcatum*: dinoflagellate cyst from the Tertiary, stained with basic fuchsin to reveal the internal microstructures, and viewed under the optical microscope (collections: A and B: UMR 8198 EEP Lille; C to E: University of Aberystwyth; F and G: Sorbonne Université, Paris; photos: A: M. Vecoli; B: D. Kroeck; C to E: F. Polette; F and G: E. Masure). Scale bars: A, C to E = 10 µm; B, F and G = 20 µm.*

Example of coccoliths

Coccolithophorides are single-celled marine algae, and are part of nanoplankton, ranging in size from a few microns to a few tens of microns. Unlike diatoms, which are siliceous, these micro-algae have a spherical skeleton made of calcium carbonate ($CaCO_3$). This carbonated structure constitutes the coccosphere (see Plates 3.21A and 3.21B) – an assembly of calcareous plates arranged in a radial fashion around the cell body, known as "coccoliths" (Bignot 2001). When the organism dies, its skeletal elements tend to fall off and build up on the sea bed (see Plates 3.21A to 3.21D). Their accumulation in the ocean depths during the Late Cretaceous gave rise to a very particular kind of carbonaceous sediment: chalk. However, coccoliths in the broad sense have lived since the Triassic and are still found today (Gardin et al. 2012). It should be noted that, in order to build their calcium carbonate skeleton, these micro-algae capture carbonic gases dissolved in seawater, thereby immobilizing those greenhouse gases. From a mineral standpoint, coccoliths are formed of calcite with low magnesium content, and can easily be fossilized in that carbonated form in the chalky sediments of the Upper Cretaceous (Püttmann and Mutterlose 2021). However, they may also be found in other fine-grained carbonated rocks – notably paper shale (e.g. the Toarcian deposit in the Paris basin), lithographic limestone (e.g. from the Mid- and Upper Jurassic), or marl (Goy et al. 1979; Noël et al. 1994a,

1994b; Clémence et al. 2015). It should be noted that in finely laminar facies with some bitumen content (for example the paper shale from the Toarcian), coccoliths may be found in abundance in the light-colored carbonaceous strata, but are rarer, or entirely absent, in the darker layers, which are more organic, and richer in framboidal pyrite. However, they may be preserved in the form of casts or imprints in pyrite or in amorphous organic material (Batten 1985). Certain authors have even pointed out that some forms of coccoliths are highly resistant to the processes of dissolution and during the diagenesis (Defrandre and Dangeard 1983; Roth 1984). In addition, as chalk is a rock that is frequently affected by silicification, it is possible to find silicified coccoliths in the cavities or powdery enclaves of flint nodules (Powell 2012).

Example of radiolaria

Radiolaria are single-celled organisms comprising a rigid capsular membrane (pseudochitin?), an endoplasm with a nucleus and various inclusions, and a vacuolar ectoplasm which secretes a siliceous skeleton (Bignot 2001; De Wever 2016). In the fossilized state, it is this skeleton which is generally preserved in a siliceous form, either as amorphous silica (opal-A), or epigenized into quartz microcrystals. More rarely, radiolaria may be preserved in epigenized form as calcite or pyrite. Opal radiolaria are found in abundance in particular rocks known as radiolarites or jaspers (cherts) (see Plates 3.21E and 3.21G to 3.21I), while radiolaria recrystallized as microquartz (see Plate 3.21F), pyrite or calcite may also be found in limestone. The way in which they are extracted from the rock will be different depending on the type of gangue in which they are contained: release using dilute hydrofluoric acid in the case of radiolarites and jaspers, or using hydrochloric acid for limestone.

Radiolaria calcitized in limestone are difficult to extract because both the microfossil and the gangue surrounding it are made of calcium carbonate, but in the same type of rocks, specimens made of pyrite can be extracted using HCl (Danelian et al. 2007).

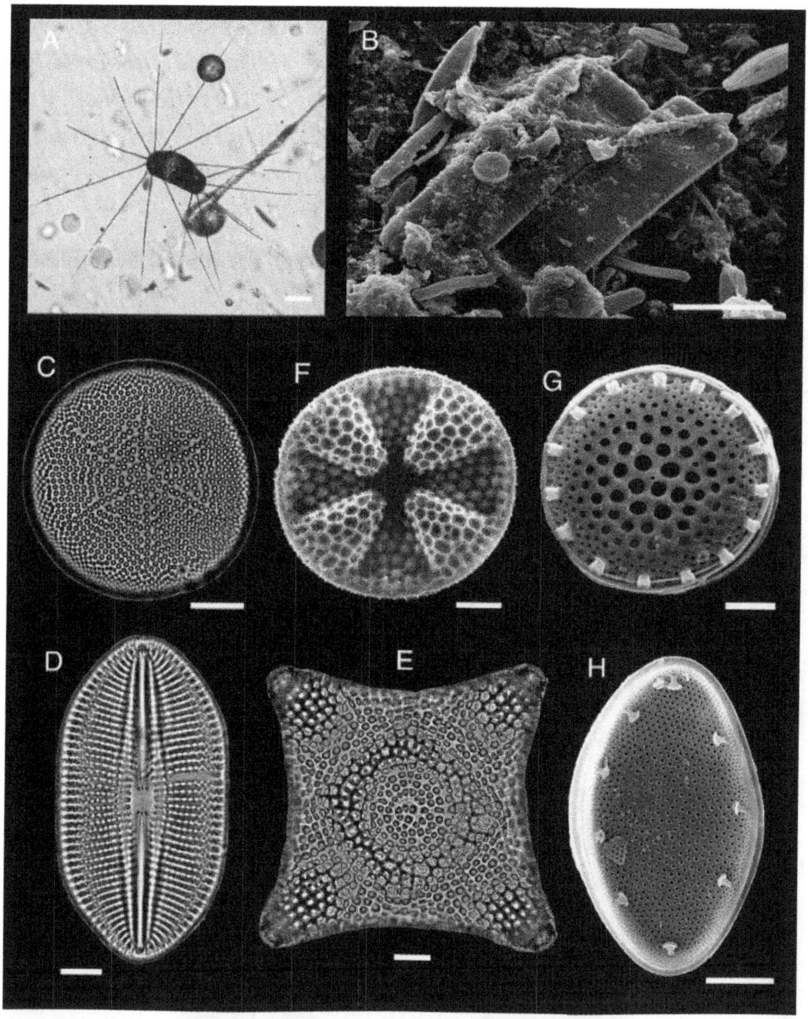

Plate 3.20. *Diatoms observed in different contexts and by different methods*

NOTE ON PLATE 3.20.– *A:* Corethron *preserved in amber, Turonian of Vendée; B: present-day diatoms encrusting the surface of an amber fragment, Cenomanian of Charente-Maritime, south-western France; C:* Actinocyclus ehrenbergii, *Messinian of Oran, Algeria; D:* Diploneis smithii, *Messinian of Oran, Algeria; E:* Triceratium balearicum,

Messinian of Oran, Algeria; F: Actinoptychus senarius *observed under SEM; G:* Thalassiosira decipiens *observed under SEM; H:* Hemidiscus cuneiformis *observed under SEM (collections: A and B: Girard-IGR; C to H: J.-P. and S. Saint-Martin-MNHN; photos: A to B: V. Girard; C to H: J.-P. and S. Saint-Martin). Scale bars: A = 50 μm; B to D, F = 10 μm; E = 20 μm; G = 2 μm.*

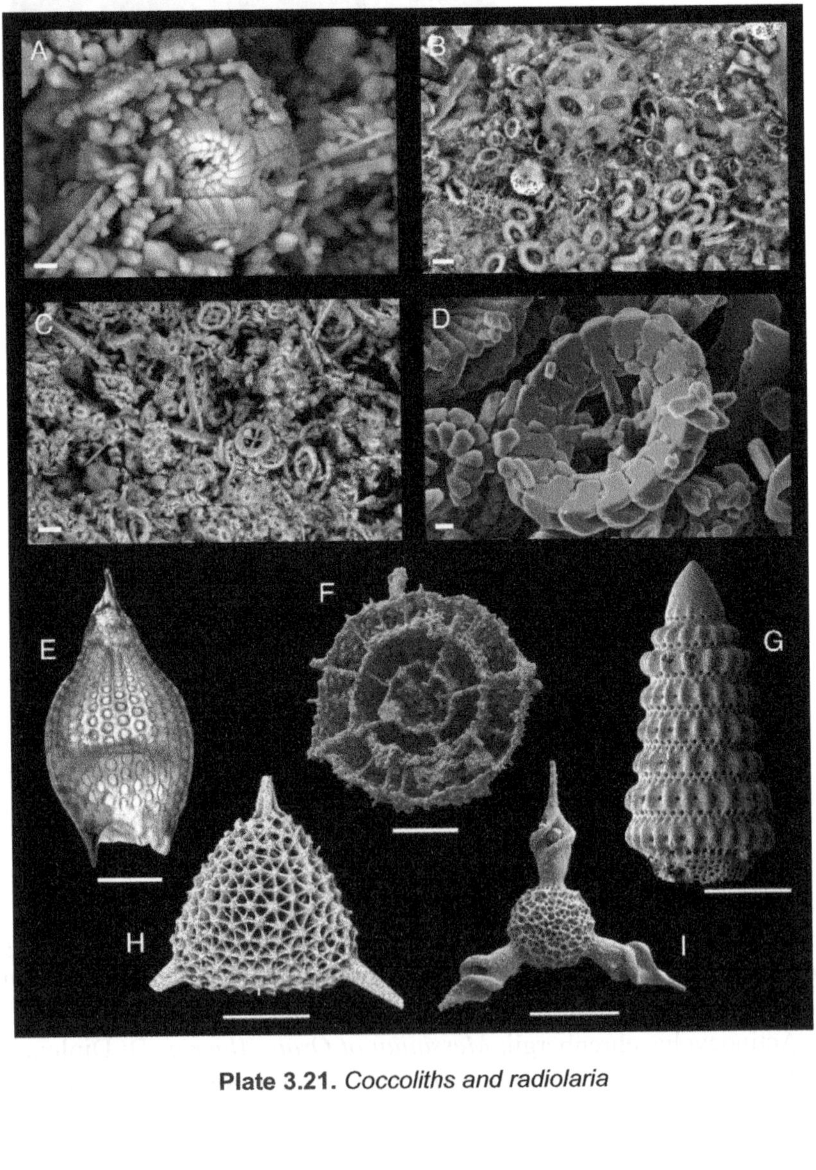

Plate 3.21. *Coccoliths and radiolaria*

NOTE ON PLATE 3.21.– *A to D: coccoliths; E to I: radiolaria. E:* Podocyrtis papalis, *Eocene, preserved in amorphous silica (opal-A) and observed under optical transmission microscopy; F:* Triplococcus *sp., Ordovician, preserved in microcrystalline quartz and observed under a scanning electron microscope (SEM); G:* Pseudodictyiomitra *sp., Upper Cretaceous (Cenomanian), Romania, observed under SEM; H:* Alievum superbum, *Upper Cretaceous (Campanian), Romania, observed under SEM; I:* Capnuchosphaera triassica, *Triassic (Carnian), Türkiye, observed under SEM (collections: A to D: Sorbonne Université; E and F: Université de Lille; G to I: MNHN, Paris; photos: A to D: S. Gardin; E to F: T. Danelian; G to I: P. DeWever). Scale bars: A = 1 μm; B = 3 μm; C = 2 μm; D = 0.3 μm; E and G = 50 μm; F and H = 100 μm; I = 200 μm.*

Example of foraminifera

Foraminifera are single-celled organisms (protozoans), most of which have a mineralized test, with the exception of certain forms, which have a purely organic envelope (chitinoid), which are rarely fossilized (Bignot 2001). The test of the mineralized species may be carbonated, when it is made by the organism itself, or biodetritic if it is made of debris taken from the surroundings.

Carbonated forms resemble essentially calcitic species, with a microgranular or a pseudofibrous structure, a shiny white porcelaneous structure, or micro- or macrocrystalline and hyaline, which are more translucent (see Plate 3.22). Calcitic species have a high potential for fossilization in limestone and marl sediments, as in these contexts the geochemical nature of these microfossils is similar to that of the surrounding sediment. However, these calcitic foraminifera are more rarely fossilized in sediments which can become "acidic contexts" due to water percolation, such as sand, sandstone or hydrophilic clay.

Note that they may be fossilized in siliceous form, notably in flints, but in this case, we are seeing the epigeny of tests which were originally carbonated (see Plate 3.22F).

As for species with an aragonitic test, the fragility of aragonite makes them particularly vulnerable to dissolution, or they may be fossilized if their test quickly recrystallizes as calcite. Forms with a biodetritic envelope, known as "agglutinative" or "agglutinating", assemble their test from grains of quartz, and particles of a calcareous or other nature, bound together by organic or carbonated secretions. Species agglutinating siliceous elements are the most resistant to dissolution and, unlike carbonated species, may be abundantly preserved in sandy or sandstone sediments. Such is the case with the large benthic agglutinative foraminifera which are the orbitolina from the mid-Cretaceous and the nummulites from the Paleogene (see Plate 3.22).

Note that in sediments subject to significant decalcification (diagenetic or by acid attack in the laboratory), notably the ligniteous clay that is so effective at preserving palynomorphs, the only remains of foraminifera found are the organic basals of their tests (Vénec-Peyré 1981; Néraudeau et al. 2002). Such foraminifera basals have even been found in amber, with these marine micro-organisms having been deposited by sea spray on resin flowing from conifers growing on the coast (Girard et al. 2008) (see Plate 3.22G).

Ostracods are microfossils, but unlike diatoms or foraminifera, which are single-celled organisms, they are minuscule metazoans with a complex anatomy, belonging to the crustaceans and therefore to the arthropods. They measure between a few millimeters and a few tens of millimeters, and their thickset bodies, with a short trunk and a developed head, is encased in a two-valve carapace. Some forms of ostracods have an organic carapace, and do not tend to be fossilized well. However, most species have calcareous valves which vary in thickness, smooth or adorned with ribs or tubercles. This carbonated nature, as is the case for foraminifera with a calcitic test, enables them to be fossilized well in limestone or marl sediments, but they are more rarely preserved in sand, sandstone and clay. Their distinguishing feature, like other arthropods, is that during the course of their lives, they slough off their carapace multiple times and grow a new one, meaning a single individual contributes multiple carapaces to the surrounding environment, thus increasing its chances of being fossilized.

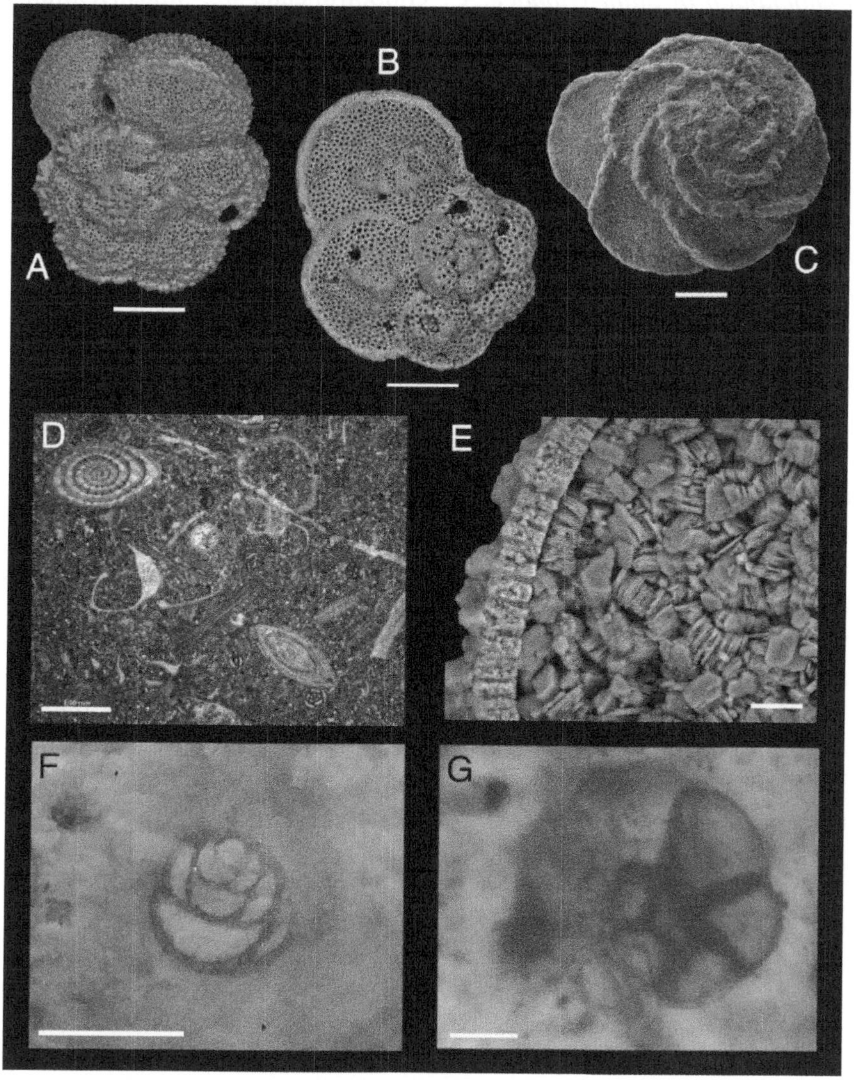

Plate 3.22. *Foraminifera viewed in 3D and in cross-section*

NOTE ON PLATE 3.22.– *A:* Praeglobotruncana rillela, *mid-Cenomanian of Saint-Ulphace, central France; B:* Rotalipora cushmani, *mid-Cenomanian of Saint-Ulphace; C:* Globotruncana *sp., Maastrichtian, Bidart; D: thin section in the Paleogene deposits of Coustouges, southern France, with sections of various foraminifera; E: sheets of clay crystals filling the hollows on a* Whiteinella, *mid-Cenomanian of Hot Springs, South Dakota, USA; F: foraminifera visible on the surface of a Campanian flint fragment, from the paleolithic site of Menez Dregan, Finistère, north-western France; G: basal of a foraminifera preserved in amber, Cenomanian of Charente-Maritime, south-western France (collections: A to E: Sorbonne Université; F: UMR Archeosciences Université de Rennes; G: IGR; photos: A to E: D. Desmares; F: A. Trébaol; G: V. Girard). Scale bars: A to C = 100 μm; D = 2.3 mm; E and G = 10 μm; F = 0.5 mm.*

Foraminifera and ostracods can be collected by washing/sieving and dissociation of the least hardened sediments (sand, marls). They can then be recovered by visual sorting, observed in three dimensions under a binocular lens, or a scanning electron microscope (see Plate 3.21A, 3.21C and 3.21E), depending on their size. When searching for them in greatly hardened or "vitreous" sediments (limestone, sandstone, flints, amber), we need to take very thin slices of rock, as these thin sections can be observed by transparency optical microscopy (see Plate 3.21D to 3.21G). However, they can sometimes be observed on polished surfaces after sawing or the natural surfaces after the breaking of these rocks (see Plate 3.21F) – notably in flints.

4

Molecular Preservation

4.1. Color fossilization

Generally, fossils of plants or animals, be they marine or terrestrial, regardless of their geological age, have lost their original colors. For example, we do not find fossilized leaves that have remained green, fossilized flowers that are still colorful, or shells from the Mesozoic or Cenozoic bearing multicolored patterns.

Plants (leaves, wood, flowers, seeds) are usually transformed into carbonaceous residues which are black or brown. Mollusk shells are whitened over time. Conversely, the white teeth of vertebrates, when they are fossilized, take on brown, black, gray or bluish tints as a result of oxidation or recrystallization.

However, there are some cases of exceptional preservation of the original colors – particularly in mollusks and crustaceans, and there are also processes by which colored patterns can be revealed on fossils which have turned white or brown.

For a color version of all the figures in this chapter, see www.iste.co.uk/neraudeau/fossilization.zip.

4.1.1. *Fossils with preserved colors*

The most frequently preserved direct colors are variants of red or purple, generally due to the presence of iron in the mineral structure of the organisms in question. The iron oxides responsible for the reddish tinges during the animal's lifetime are altered little by fossilization, and remain in place on the fossil, albeit probably with certain variations between the hues of the original colors and those of their fossilized remains.

In bivalve mollusks, which live in today's seas in a wide variety of colors, notably in pectinids (*Pecten* or *Chlamys* which are yellow, orange, red, pink, purple or brown), cases of direct preservation of colored patterns are extremely rare. We can point to a remarkable case in the oysters of the mid-Cretaceous in France – the exogyra *Rhynchostreon suborbiculatum* from the Cenomanian of Sarthe (central western France, see Plates 4.1E to 4.1G) and, to a lesser extent, the *Pycnodonte biauriculatum* associated with the former (see Plate 4.1H).

In gastropod mollusks, preservation of colors mainly happens for fossils from the Tertiary. Rather than the conservation of shells with a uniform color, what tends to be preserved is the instead geometric colored patterns that ornament certain shells, such as points, flames and stripes (see Plates 4.1A to 4.1D).

In crustaceans, which are also highly colorful in present-day seas, notably in decapods (crabs or lobsters ranging from bright blue to vibrant red), the fossilization of species with the preservation of colors is just as rare as it is in bivalves. A fairly esthetically pleasing example, though discrete, is that of the cirripedes in barnacles, which are found in abundance in the Miocene faluns of Anjou, Britanny and Touraine. Indeed, the "giant" species *Megabalanus tintinnabulum* is often found with pink, carmin or purplish-blue ornamentation, either with a network of colored spots alternating with white spots, or with a more general spread of color (see Plate 4.1J).

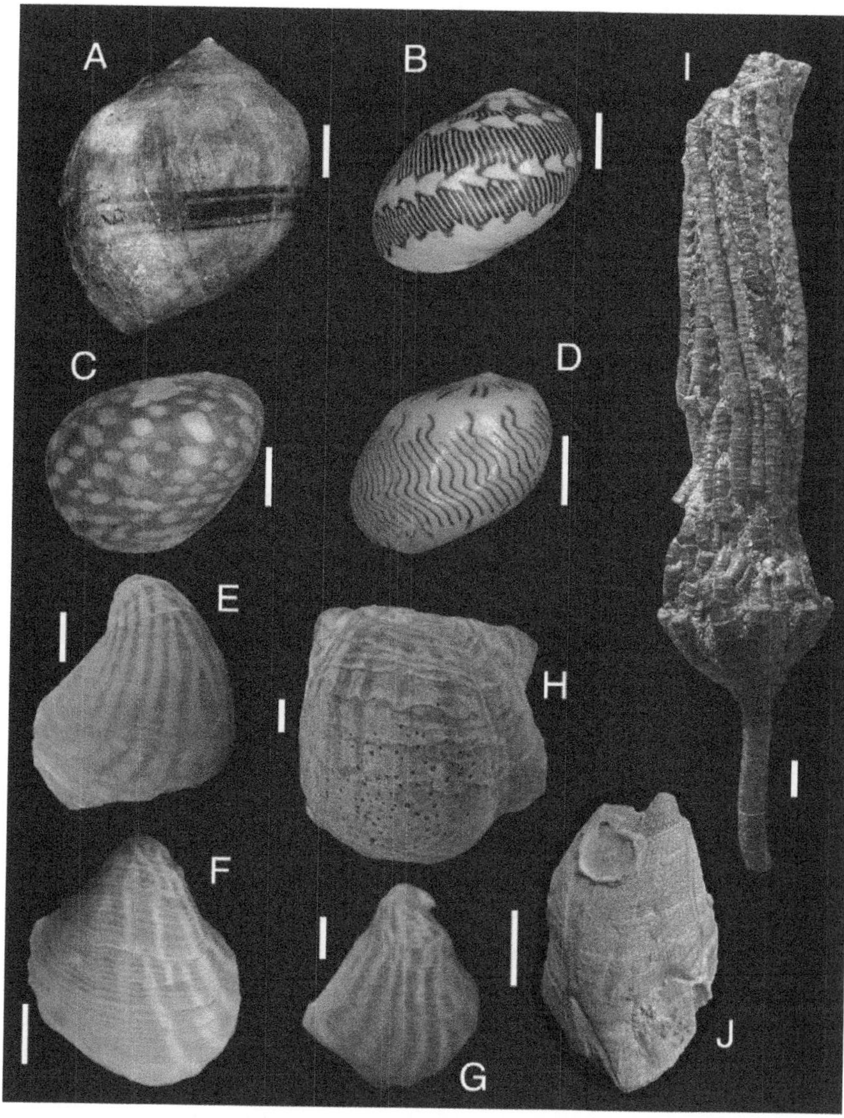

Plate 4.1. *Colors preserved on marine fossils*

NOTE ON PLATE 4.1.– *A to D: Cenozoic gastropods with their colored patterns preserved. A:* Ampullina perusta, *Eocene of Italy; B:* Clithon inequidentatum, *Eocene of France; C:* Nerita symondsi, *Eocene of*

France; D: Smaragdia merignacensis, *Miocene of France; E to H: Cretaceous oysters from the Cenomanian of Sarthe (central western France), with their colored patterns preserved; I: crinoid* Millericrinus, *from the Upper Jurassic of Angoulins (Charente-Maritime, western France), with its purple color preserved; J: cirriped crustacean (barnacle) (balane), mid-Miocene of Tréfumel (Ille-et-Vilaine, north-western France) with a pale pink or carmine matrix preserved (collections: A to D: MNHN, Paris; E to H: Musée Vert, Le Mans; I: Muséum d'Histoire naturelle de La Rochelle; J: IGR; photos: A to D: MNHN; E to H: N. Morel; I: R. Vullo; J: D. Néraudeau). Scale bars: A to D = 2 mm; E to I = 1 cm; J = 5 mm.*

Among echinoderms, colored fossils are found quite frequently, but almost only in pedunculate crinoids from the uppermost Jurassic. Thus, in the perirecifal facies of the Oxfordian and Kimmeridgian, notably in Charentes (south-western France), we find calyxes and stems of *Apiocrinus* colored dark blue-purple, and comparable skeletal elements of *Angulocrinus*, of a clearer shade of purple, closer to pink (see Plate 4.1I).

Many groups of insects have colorful patterns on their wings, in multi-form and multiple spots, eyespots, or stripes. Even if the truly original color is not preserved in fossils, we can fairly often observe the size and shape of these patterns on fossilized wings (and less frequently on the bodies) in laminar sediments from various ages: polyneoptera from the Triassic of South Korea (Jouault et al. 2022b) (see Plate 4.2A), neuroptera from the Jurassic of China (Jouault et al. 2022a) (see Plate 4.2E), odonata from the Jurassic of Germany (Jouault et al. 2022d) (see Plate 4.2B), orthoptera from the Lower Cretaceous of Brazil (Nel and Jouault 2022) (see Plate 4.2D) and mecoptera from the Upper Miocene of Cantal, central southern France (see Plate 4.2C). Colored patterns may also still be visible on insects preserved in amber, such as the spots which pepper the wings of certain neuroptera from the Cretaceous of Myanmar (see Plate 4.2F) (Jouault et al. 2022c). Most often, though, it is the general coloration of the thorax which is preserved, with bluish or greenish metallic colors (Cai et al. 2020).

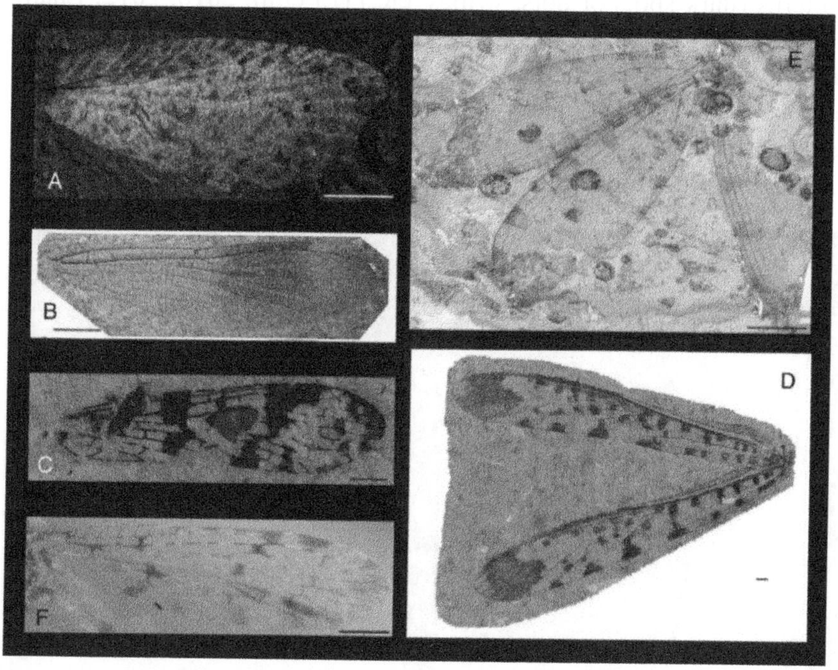

Plate 4.2. *Preserved colors on fossilized insects*

NOTE ON PLATE 4.2.– *Fossils of insects with colored patterns preserved, either as an impression in laminar sediments (A to E) or in amber (F). A: wing of a polyneoptera, Upper Triassic, South Korea; B: wing of an odonata, Lower Jurassic, Germany; C: wing of a mecoptera, Upper Miocene of Cantal, France; D: wings of a neuroptera, Lower Cretaceous of Brazil; E: wing of a neuroptera, mid- to Upper Jurassic of China; F: wing of a neuroptera in amber, mid-Cretaceous of Myanmar (collections: A: Kongju National University; B: Beijing Institute of Geology and Paleontology; C: H. Tischlinger; D to F: IGR; photos: A: G.S. Nam; B to F: C. Jouault). Scale bars: A and C = 2 mm; B = 5 mm; D = 1 mm; E = 10 mm; F = 1 mm.*

Finally, in vertebrates, it is primarily feathers which, as they are fossilized, may retain not their original colors but rather their colored patterns (dark spots or dots, stripes, etc.) (Li et al. 2010, 2012; Zhang et al. 2010). This type of preservation is seen mainly in fine sediments in the structure of leaflets, such as lithographic limestones, and laminar limestone even more so. Such is the case, notably, in the fine layered limestone from the Oligocene of Céreste, in the Lubéron Geological Reserve (southern France, see Plates 4.3A and 4.3B).

Plate 4.3. *Colors preserved on fossilized feathers*

NOTE ON PLATE 4.3.– *Fossilized feathers with preserved colored patterns, found in the layered limestone from the Oligocene of Céreste, Lubéron, southern France. A and B: feathers whose extremities have darker tinges (collection: Lubéron Geological Reserve; photos: A: C. Jouault; B: R. Zurbach). Scale bars: 1 cm.*

4.1.2. *Fossils where the colors are absent, but can be revealed*

Though bivalves and gastropods are very frequently fossilized in the form of internal casts, for epochs prior to the Cenozoic, those of the Tertiary, and primarily of the Eocene, are preserved with their shells. However, these carbonated structures have generally been whitened by diagenesis and their lengthy stay in ancient sediments, tens of millions of years old. Despite this, it has been shown that often, their original colored patterns can be revealed under ultraviolet lighting and after a special treatment with chlorinated water (Caze et al. 2011; Merle 2022).

What appears under UV is not truly the original colors, but rather the contrast between the zones colored differently on the animal's shell during its lifetime. Thus, we can see concentric patterns, checkerboard patterns, flame shapes, or radial or speckled strips. The effect of the UV rays expresses these ornamentations with shades of blue, pink or red, but the original colors may be much less vibrant or indeed completely different. In a somewhat similar way, in rare cases, we can also reveal the colored patterns on fossilized feathers – for example, stripes and splotches – and indeed we can gain an idea of the colors, more towards the brown/red area of the spectrum or more towards the gray/black, depending on whether iron or manganese, for example, are present in the organic residues.

By studying the morphology of the melanosomes of the feather, we can also glean some information about its color. For example, analyzing the melanosomes of the small feathered theropod *Microraptor* (Barremian, China) revealed that its plumage was iridescent (for the purposes of seduction/parading) and primarily blue-black (Li et al. 2012).

4.2. Fossilization of the original chemical compositions

Example of the chemical formulas of plant resins

Fossilized resins and gums are complex organic compounds with long carbon chains. Each plant family has a molecular signature of these resins, which tends to withstand fossilization reasonably well. Thus, we can analyze ambers and copals to determine which types of trees produced them, at least on the scale of the major groups. Thus, for Cretaceous ambers, it is possible to differentiate the families of the Araucariaceae, Cheirolepidiaceae and Cupressaceae. Their molecular signature, through amber, can give us an idea of the "resiniferous" vegetation, and thus the landscape which was home to the animals trapped in the fossilized resin – notably insects and arachnids.

Example of air or water bubbles

Theoretically, air or water bubbles have little likelihood of being fossilized. However, two contexts are favorable to their preservation: meso-Cenozoic ambers and Pleistocene ice. In numerous ambers and copals, we can observe bubbles of air, trapped when the resin was flowing on the tree trunks or branches. Unpublished analyses (D. Néraudeau pers. obs.) of the Cretaceous ambers from Charentes (south-western France) showed that the original composition of the air contained in these bubbles had not been altered significantly, either by diagenesis or by the passage of time.

In reality, there may be a loss of fluid when the amber is finely cracked but, conversely, we practically never see pollution of the bubbles by the atmosphere surrounding the amber after it is excavated. Hence, these bubbles in amber constitute a direct and exceptional record of the atmosphere of the past.

A similar approach is often applied to the air bubbles contained in polar ice, which date from much more recently, and essentially document the ancient atmospheres of the Quaternary (Masson-Delmotte and Chapellaz 2002; Raynaud and Lorius 2004).

Similarly, though more rarely than air bubbles, amber may sometimes be found to contain drops of water. Usually this is water that was encapsulated by the resin when it fell into a waterway, but may also be drops of seawater captured in the resin as it fell from a conifer in an immediately coastal context.

4.3. Fossilized DNA

DNA is a fragile molecule which can be oxidized, hydrolyzed and fragmented quickly, in terms of the geological timescale. It is only possible for it to be preserved in very specific sedimentary and climatic contexts, such as in "frozen" form in the permafrost, or in isothermal sedimentary infills of prehistoric caves. Its preservation over a long period of time, over a million years, is also unlikely, or at least highly controversial (Willerslev and Cooper 2005).

The DNA molecule cannot be fossilized in just any remains of an organism; to a large extent, hair, teeth, bones and excrement (coprolites) are the remains of vertebrates which are most promising from the point of view of paleogenetics (Willerslev and Cooper 2005; Geigl 2008; Campos et al. 2012). For plants, subfossil seeds (from the Pleistocene or Holocene) offer good potential. As for micro-organisms, they are notably preserved in ancient ice, from which they can be sampled by means of a drill core. On the other hand, fossilized resins, such as copal and amber, do not lend themselves well to DNA preservation, contrary to the myth that was developed by *Jurassic Park* (Penney 2013).

Finally, sedimentary or environmental DNA may be preserved in ancient soils, where over long periods of time multiple species have deposited an accumulation of debris of skin, hair, excrement and body fluids, combined with various plant or microbial elements (Willerslev and Cooper 2005).

After spectacular results, which were soon invalidated during the 1990s and 2000s, where researchers believed they could recover the genetic material of insects trapped in amber, or of dinosaurs, techniques and safeguards have since been improving steadily (Poinar and Stankiewicz 1999). Research into fossil DNA has been expanding since the early 2000s, primarily targeting large Quaternary mammals such as Neolithic aurochs (Geigl 2008) and the earliest domesticated horses (Orlando et al. 2013), wolves (see Plate 4.4A) (Bergström et al. 2022), mammoths (see Plate 4.4B) and indeed ancient humans, with Neanderthal man being the iconic scientific support (Krings et al. 1997, 1999; Green et al. 2009; Prüfer et al. 2014).

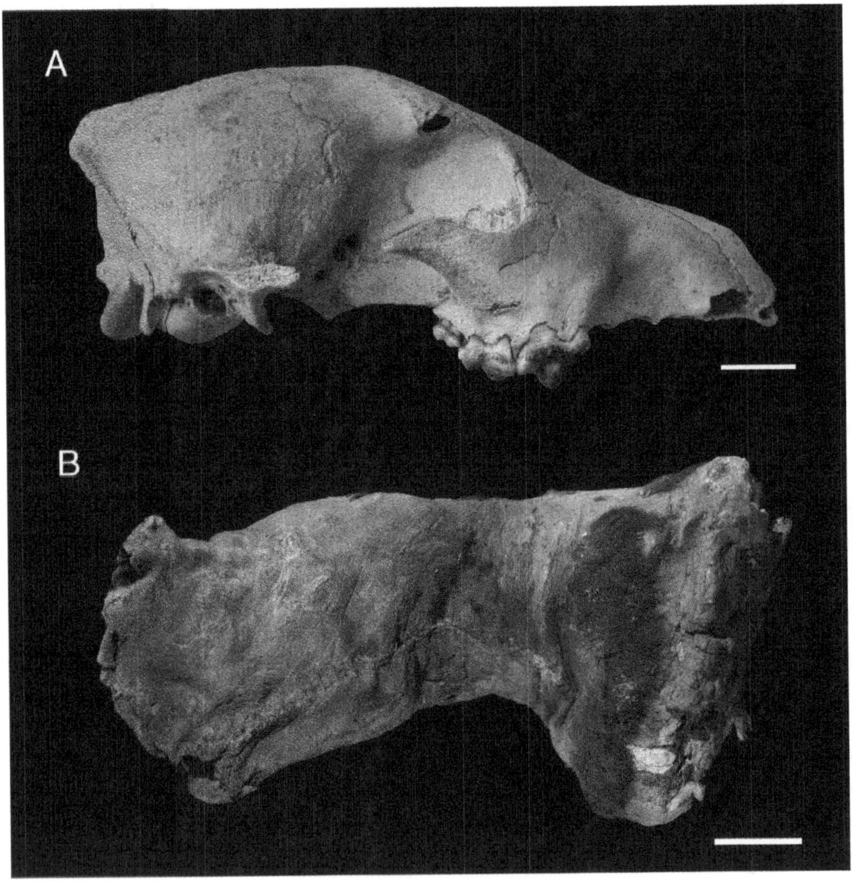

Plate 4.4. *The remains of Pleistocene mammals which can be used to support the quest to recover fossilized DNA*

NOTE ON PLATE 4.4.– *The remains of mammals from the ancient Quaternary whose fossils include teeth or hair which can serve as a starting point when looking for fossilized DNA. A: the skull of* Canis *sp., Pleistocene deposits from the River Sélune, Manche, north-western France; B: the tip of the foot from a Pleistocene wooly mammoth (collections: A: A. Stuer-IGR; B: MNHN, Paris; photos: A: D. Néraudeau; B: P. Tassy). Scale bars: A = 1 cm; B = 15 cm.*

With respect to wolves, subfossil skeletons dating from the Pleistocene or Holocene have been found at various sites in different parts of the world. Using these markers, it is possible to trace the history that led from wild wolves to the domesticated forms that became dogs (see Figure 4.1).

The excellent preservation of the bones, crania, and in particular the teeth, in anoxic clay-sand sediments in a temperate context or in permafrost, is an advantage when it comes to finding fossilized DNA (Bergström et al. 2022).

In France, the remains of ancient canids have been found in various deposits in the north of the country (Auguste 2009) – in particular, in the Sélune estuary, in the Manche Département (see Plate 4.2A).

Frozen mammoths have been found over the course of more than a century in the permafrost and the Pleistocene ice of Siberia. Since the end of the 20th Century, the idea of cloning mammoths has taken root in the minds of scientists in various countries, postulating that fossilized DNA was still preserved in good condition in these "cryogenized" remains from the Early Quaternary.

Though nobody has gone so far as to actually clone a mammoth, frozen mammoth DNA has frequently been analyzed in order to reconstruct the phylogenetic tree of these proboscideans – particularly the relationships between African or Asian elephants and wooly mammoths (Barriel et al. 1999; Van der Valk et al. 2021).

From a museographic standpoint, more or less complete remains of mammoths from Siberia, which still had hair on their "mummified" skin, are exhibited in numerous museums all over the world, such as the Lyakhov mammoth foot, exhibited at the Muséum national d'histoire naturelle de Paris (see Plate 4.2B).

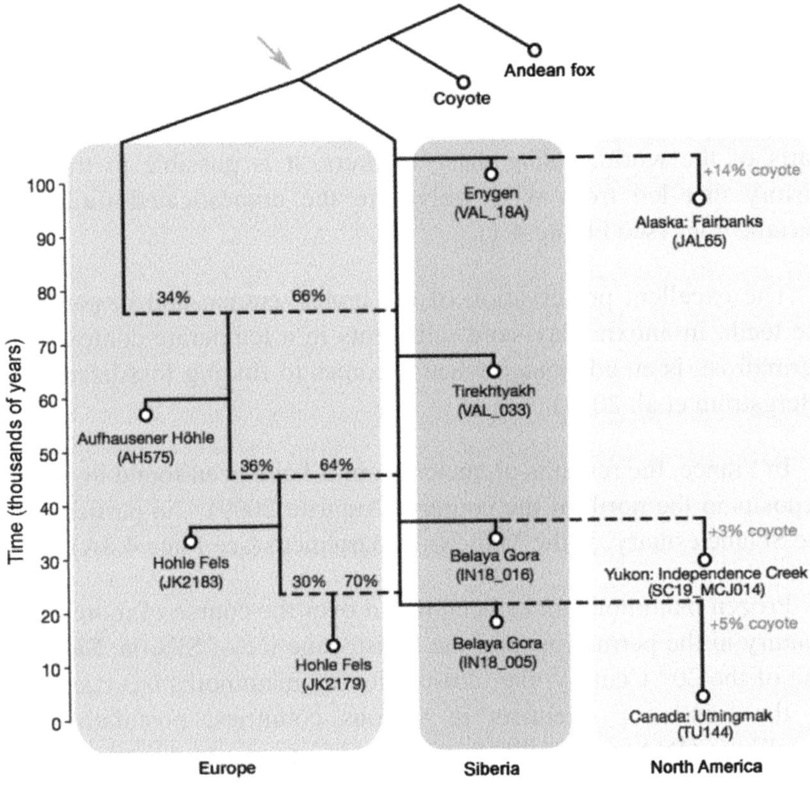

Figure 4.1. *Wolf phylogeny*

NOTE ON FIGURE 4.1.– *Diagram summarizing 100,000 years of history of the populations of gray wolves,* Canis lupus. *The phylogeny is based on DNA analysis on the genome of 72 ancient Holocene or*

Pleistocene wolves from Europe, Siberia and North America. This paleogenetic study suggests a dichotomy in the ancestors of dogs, potentially resulting from the domestication of two source populations of wolves, but with interconnections between the populations being maintained throughout the Pleistocene (modified from Bergström et al. (2022)).

In 2022, the project to sequence the genome of Neanderthal man won a Nobel Prize in Medicine for the Swedish geneticist Svante Pääbo, of the Max Planck Institute for Evolutionary Anthropology in Leipzig. A pioneer of paleogenetics, he began his work in 1997 by decoding the mitochondrial DNA (which sits outside the cell nucleus) of this species of prehistoric humans. Then, in 2010, he published the complete genome, and demonstrated interbreeding between Neanderthals and modern humans (*Homo neanderthalensis* shares 2 to 4% of genes with *Homo sapiens* populations today) (Kuhlwilm et al. 2016). By studying a human phalange from the Denisova Cave in Russia, in 2010 he also proved the existence of another human species which coexisted with Neanderthals: *Homo denisovensis* (Reich et al. 2010).

In the late 20th Century, the fossil DNA found in the hair follicles or teeth roots of frozen mammoths in the Siberian permafrost was used to prove the genetic proximity of this fossil proboscidean with its modern-day cousin the Asian elephant; African elephants are further removed, genetically, from their mammoth ancestors (Barriel et al. 1999). This type of research on ancient DNA has not only allowed scientists to establish the phylogeny of wooly mammoths, but also to shed light on their polymorphism (Barnes et al. 2007; Workman et al. 2011). In addition, the ice cores taken from the permafrost in Klondike (USA) were rich in environmental DNA. These genetic microfossils come from excrement or body fluids (feces, urine, sweat, saliva) and the phanera (hair or feathers) deposited or lost in the environment by the animals. Though environmental DNA is greatly outweighed (over 99%) by that of fungal organisms, it is possible to isolate DNA from animals, such as mammoths, bison or ancient horses, and thus paint a picture of the fauna at the time when the soil was formed. Thanks to environmental DNA, we can pinpoint an extinct species as being present in a more recent epoch than indicated by the latest known bone fossils: evidence of woolly mammoths in Klondike dates from potentially 5,700 years ago, instead of the 7,000 years indicated by bone finds.

Conclusion

Regardless of the processes involved, fossilization is always an exceptional phenomenon – the result of diverse and sometimes mutually antagonistic environmental factors, such as the respective rates of putrefaction, burial and mineralization.

The most common types of fossilization essentially result from the organisms' rapid burial, meaning they are protected from oxidative action, but also from carrion feeders and certain decomposers. Various fossilization processes may then occur, depending on whether or not sediment penetrates the cavities or porosity of the organism, and whether the original mineralized structures (the shell, the test, etc.) are dissolved or not. This results in different types of preservation of the endo- or exoskeletal parts (shells, teeth and bones): internal casts, imprints and replicas.

The most exceptional means of fossilization are those where the action of decomposers has been so limited that the least mineralized and most fragile parts (muscles, phanera, etc.) have left an imprint in the sediment, or part of their organic components. In certain cases, the preservation of the histological microstructures is such that the cell walls, nuclei and even the chromosomes that they contain may be fossilized. These particular fossilization conditions may give rise to exceptional fossiliferous deposits – conservation Lagerstätten.

Special, and even spectacular, fossils, may result from mineralizations caused by percolation, into the sediment, of fluids (alteritic, hydrothermal) enriched with various minerals. Such mineralization occurs either at the surface of the organisms or in their internal cavities (crystalline outgrowths, geodes), or infiltrate their fine porosity. The exogenous mineral is then added (addition) to the skeletal framework or the cavities of the original organism, and can even substitute it (epigeny, pseudomorphosis), either in part or in full (in the latter case, forming a replica). The most frequent cases of mineralization are associated with calcite, silica or pyrite, but remarkable cases may involve phosphates, hydroxides or molecules that give rise to precious or semi-precious minerals (such as opal and emerald).

However, *complete* preservation of an organism has never been seen, if only because the fluids (gases, blood, sap, etc.) are modified, and then generally lost, during the processes of putrefaction, compaction, desiccation and filling, even before the anatomical elements are – quite literally – set in stone by mineralization.

Because of its complex geological history, and the great lithological and stratigraphic diversity of its sedimentary formations, the French territory offers a remarkable concentration of exceptional fossiliferous deposits. These Lagerstätten fully illustrate the multitude of mechanisms and cases of fossilization, from the most common to the most exceptional.

References

Alimen, H. (1947). Présence de foraminifères dans la Meulière de Beauce. *Bull. Soc. Géol. France*, 17(7–9), 445–452.

Allain, R., Vullo, R., Rozada, L. et al. (2022). Vertebrate paleobiodiversity of the Early Cretaceous (Berriasian) Angeac-Charente Lagerstätte (southwestern France): Implications for continental faunal turnover at the J/K boundary. *Geodiversitas*, 44(25), 683–752.

Arouri, K.R., Greenwood, P.F., Walter, M.R. (2000). Biological affinities of Neoproterozoic acritarchs from Australia: Microscopic and chemical characterization. *Organ. Geochem.*, 31(1), 75–89.

Audo, D. and Charbonnier, S. (2012). Late Cretaceous crest-bearing shrimps from the Sahel Alma Lagerstätte of Lebanon. *Acta Palaeont. Pol.*, 58(2), 335–349.

Audo, D., Williams, M., Charbonnier, S., Schweigert, G. (2017). *Gabaleryon*, a new genus of widespread early Toarcian polychelidan lobsters. *J. Syst. Palaeontol.*, 15(3), 205–222.

Auguste, P. (2009). Évolution des peuplements mammaliens en Europe du Nord-Ouest durant le Pléistocène moyen et supérieur. Le cas de la France septentrionale. *Quaternaire*, 20(4), 527–550.

Azar, D., Gèze, R., Acra, F. (2010). Lebanese amber. In *Biodiversity of Fossils in Amber from the Major World Deposits*, Penney, D. (ed.). Siri Scientific Press, Manchester.

Barailler, B. (2023). Fleurs d'opale et péridotites du volcan des Baumes Hérault. *Planet Terre* [Online]. Available at: https://planet-terre.ens-lyon.fr/ressource/opale-peridotite-Baumes.xml.

Barnes, I., Shapiro, B., Lister, A., Kuznetsova, T., Sher, A., Guthrie, D., Thomas, M.G. (2007). Genetic structure and extinction of the woolly mammoth, *Mammuthus primigenius*. *Curr. Biol.*, 17, 1072–7075.

Barriel, V., Thuet, E., Tassy, P. (1999). Molecular phylogeny of Elephantidae. Extreme divergence of the extant forest African elephant. *C.R. Acad. Sci. Paris*, 322, 447–454.

Batten, D. (1985). Coccolith moulds in sedimentary organic matter and their use in palynofacies analysis. *J. Micropalaeontol.*, 4(2), 111–116.

Baudin, F., Tribovilard, N., Trichet, J. (2017). *Géologie de la matière organique*, 2nd edition. EDP Sciences, Paris.

Bercovici, A. and Broutin, J. (2008). La flore autunienne du site de l'étang de Martenet (Blanzy–Montceau-les-Mines, Bourgogne, France) : description, étude taphonomique et implications paléoenvironnementales. *C.R. Palevol*, 7(1), 1–16.

Bergström, A., Stanton, D.W., Taron, U.H. et al. (2022). Grey wolf genomic history reveals a dual ancestry of dogs. *Nature*, 607(313).

Bernier, P., Gaillard, C., Gall, J.C., Barale, G., Bourseau, J.-P., Buffetaut, É., Wenz, S. (1991). Morphogenetic impact of microbial mats on surface structures of Kimmeridgian micritic limestones (Cerin, France). *Sedimentology*, 38(1), 127–136.

Béthoux, O. (2008). The insect fauna from the Permian of Lodève (Hérault, France): State of the art and perspectives. *J. Iber. Geol.*, 34(1), 109–113.

Béthoux, O. and Nel, A. (2003). Révision de *Protagion audouini* Brongniart, 1893, du Carbonifère supérieur (Palaeoptera). *Bull. Soc. Entomol. Fr.*, 108(3), 237–244.

Bidaud, C. (1998). Biodégradation des hydrocarbures aromatiques polycycliques. Approche microbiologique et application au traitement d'un sol pollué. PhD Thesis, Ecole Normale Supérieure des Mines de Saint-Etienne, Saint-Etienne.

Bieńkowska-Wasiluk, M., Uchman, A., Jurkowska, A., Świerczewska-Gladysz, E. (2015). The trace fossil *Lepidenteron lewesiensis*: A taphonomic window on diversity of Late Cretaceous fishes. *Paläont. Zeit.*, 89(4), 795–806.

Bignot, G. (1982). *Micropaléontologie : les différents groupes de microfossiles, exploitation paléobiologique et géologique*. Dunod, Paris.

Bignot, G. (2001). *Introduction à la Micropaléontologie*. Gordon & Breach Science Publishers, Milton Park.

Billon, G. (2001). Géochimie des métaux et du soufre dans les sédiments des estuaires de la Seine et de l'Authie. PhD Thesis, Université de Lille, Lille.

Bomfleur, B., McLoughlin, S., Vajda, V. (2014). Fossilized nuclei and chromosomes reveal 180 million years of genomic stasis in royal ferns. *Science*, 343, 1376–1377.

Bomou, B., Suan, G., Schlögl, J. et al. (2021). The palaeoenvironmental context of Toarcian vertebrate-yielding shales of southern France (Hérault) [Online]. Available at: https://hal.science/hal-03384272/file/Bomou%20et%20al.,%202021_hal.pdf.

Borkow, P. and Babcock, L.E. (2003). Turning Pyrite concretions outside-in: Role of biofilms in pyritization of fossils. *The Sedim. Rec.*, 1, 4–7.

Bouat, L. (2023). Les minéralisations hydrothermales à l'interface socle-couverture de Vendée : origines et circulation des fluides, marqueurs de la dynamique extensive mésozoïque. PhD Thesis, Université de Le Mans, Le Mans.

Bourel, B., Moreau, J.-D., Trincal, V., Walter-Simonnet, A.-V. (2016). Fossiliferous Holocene tufa of Mende (Lozère, southern France): Implication for the Atlantic vegetation of the Causses Basin. *Bull. Soc. Géol. Fr.*, 187(4–5), 225–235.

Brachaniec, T., Leko, K., Wieczorek, A. (2016). Regurgitalites from the Silurian of Holy Cross Mountains, southern Poland. *N. Jahrb. Geol. Paläont. Abh.*, 280(3), 331–334.

Branquet, Y., Laumonier, B., Cheilletz, A., Giuliani, G. (1999). Emeralds in the Eastern Cordillera of Colombia: Two tectonics settings for one mineralization. *Geology*, 27(7), 597–600.

Bréhéret, J.-G., Hanzo, M., El Albani, A., Iatzoura, A. (2004). Impact de la vie benthique sur la genèse de nodules calcaires dans les black shales. *C.R. Géoscience*, 336(15), 1355–1362.

Breton, G. (1992). Les Goniasteridae (Asteroidea, Echinodermata) jurassiques et crétacés de France. PhD Thesis, Université de Caen, Caen.

Breton, G. (2011). Le Gault du Perthois : terriers, phosphates et petites crottes. *Bull. Assoc. Géol. Auboise*, 31, 3–66.

Breton, G. and Tostain, F. (2005). Les microorganismes de l'ambre cénomanien d'Écommoy (Sarthe, France). *C.R. Palevol*, 4, 31–46.

Breton, G., Wisshak, M., Néraudeau, D., Morel, N. (2017). Parasitic gastropod bioerosion trace fossil on Cenomanian oysters from Le Mans, France, and its ichnologic and taphonomic context. *Acta Palaeont. Pol.*, 62(1), 45–57.

Broutin, J., Chateauneuf, J.J., Galtier, J., Ronchi, A. (1999). L'Autunien d'Autun reste-t-il une référence pour les dépôts continentaux du Permien inférieur d'Europe ? Apport des données paléobotaniques. *Géologie de la France*, 2, 17–31.

Buffetaut, É., Pouit, D., Rigollet, L., Archambeau, J.-P. (1989). Poissons et reptiles continentaux du Purbeckien de la région de Cognac (Charente). *Bull. Soc. Géol. Fr.*, 5, 1065–1069.

Cai, C., Tihelka, E., Pan, Y., Yin, Z., Jiang, R., Xia, F., Huang, D. (2020). Structural colours in diverse Mesozoic insects. *Proc. R. Soc. B.*, 287, 20200301.

Campos, P.F., Craig, O.E., Turner-Walker, G., Peacock, E., Willerslev, E., Thomas, M., Gilbert, P. (2012). DNA in ancient bone. Where is it located and how should we extract it? *Ann. Anat.*, 194, 7–16.

Carcinelli, S., Ferrari, G.A., Magaldi, D. (1989). Les accumulations carbonatées de type "calcrete" dans les sols et formations superficielles d'Italie méridionale. *Méditerranée*, 68, 51–59.

Cayeux, L. (1930). Patine des silex de la Craie. *Bull. Soc. fr. Minér.*, 53(1–6) 60–72.

Caze, B., Merle, D., Saint Martin, J.-P., Pacaud, J.M. (2011). Contribution of residual colour patterns to the species characterization of Caenozoic molluscs (Gastropoda, Bivalvia). *C.R. Palevol*, 10(2–3), 171–179.

Channing, A. and Edwards, D. (2009). Yellowstone hot spring environments and paleo-ecophysiology of Rhynie chert plants: Towards a synthesis. *Plant Ecol. Div.*, 2, 111–143.

Charbonnier, S. (2009). *Le lagerstätte de La Voulte. Un environnement bathyal au Jurassique*. Muséum national d'Histoire naturelle, Paris.

Charbonnier, S., Audo, D., Caze, B., Biot, V. (2014). The La Voulte-sur-Rhône Lagerstätte (Middle Jurassic, France). *C.R. Palevol*, 13(5), 369–381.

Chukanov, N.V. (2005). Minerals of the Kerch iron-ore basin. *Mineralogical Almanakh*, 8, 1–113.

Clémence, M.-É., Gardin, S., Bartolini, A. (2015). New insights in the pattern and timing of the Early Jurassic calcareous nannofossil crisis. *Palaeogeogr., Palaeoclimat., Palaeoecol.*, 427, 100–108.

Colin, J.-P., El Albani, A., Fürsich, F.T., Martin-Closas, C., Mazin, J.-M., Billon-Bruyat, J.P. (2004). Le gisement "Purbeckien" de vertébrés de Cherves-de-Cognac, Charente (SW France) : nouvelles données biostratigraphiques. *C.R. Palevol*, 3, 9–16.

Colin, J.-P., Néraudeau, D., Perrichot, V. (2011). Termite coprolites (Insecta: Isoptera) from the Cretaceous of western France: A palaeoecological insight. *Rev. Micropaléont.*, 54, 129–139.

Collet, L.W. (1906). Les concrétions phosphatées et la glauconie des mers actuelles. *Eclogae Geol. Helv.*, 9, 118–119.

Coquel-Poussy, N. (2013). Recherches paléontologiques dans le Toarcien basal de la Lozère. *Fossiles*, 15, 5–15.

Danelian, T., Lahsini, S., De Rafélis, M. (2007). Upper Jurassic Radiolaria from the Vocontian basin of SE France. *Eclogae geol. Hev.*, 99(1), S35–S47.

De Franceschi, D. and De Ploëg, G. (2003). Origine de l'ambre des faciès sparnaciens (Éocène inférieur) du Bassin de Paris : le bois de l'arbre producteur. *Geodiversitas*, 25(4), 633–647.

De Saporta, G. (1868). Prodrome d'une flore fossile des travertins anciens de Sézanne. *Mém. Soc. Géol. Fr.*, 8(3).

De Wever, P. (2016). *Merveilleux microfossiles. Bâtisseurs, Chronomètres, Architectes*. Biotope Éditions, Mèze.

Defaut, B., Burg, J.-P., Leyreloup, A.F., Romney, F., Fuchs, Y., Alabouvette, B., Lefavrais-Raymond, A. (1990) *Notice explicative, carte géologique au 1/50 000 de la France, feuille de Séverac-le-Château (865)*. Bureau de Recherches Géologiques et Minières, Orléans.

Deflandre, G. and Dangeard, L. (1983). *Schyzosphaerella,* un nouveau microfossile méconnu du Jurassique moyen et supérieur. *C.R. Acad. Sci. Paris*, 207, 1115–1120.

Demathieu, G., Gand, G., Sciau, J., Freytet P. (2002). Les traces de pas de dinosaures et autres archosaures du Lias inférieur des Grands Causses, Sud de la France. *Palaeovertebrata*, 31, 1–143.

Devouard, B. and Bornet, R. (2015). Vous avez dit lussatite ? *Le Règne Minéral*, Special Issue, 21, 51–54.

Diot, H., Femenias, O., Moreau, C., Gaufriau, A., Roy, C., Karnay, G. (2007). *Notice explicative de la feuille Fontenay-le-Comte à 1/50 000*. Bureau de Recherches Géologiques et Minières, Orléans.

Dornbos, S.E. (2010). Phosphatization through the Phanerozoic. In *Taphonomy, Second Edition: Process and Bias Through Time*, Allison, P.A., Bottjer, D.J. (eds). Springer, Dordrecht.

Dunlop, J.A. and Garwood, R.J. (2018). Terrestrial invertebrates in the Rhynie chert ecosystem. *Philosophical Trans. Roy. Soc. B, Biol. Sci.*, 373(1739), 20160493.

Ellenberger, P. (1984). Données complémentaires sur la zonation ichnologique du Permien du Midi de la France : bassins de Lodève, Saint-Affrique et Rodez. *C.R. Acad. Sci. Paris*, 299, 581–586.

Fauconnier, D. and Masure, E. (2004). *Les dinoflagellés fossiles. Guide pratique de détermination. Les genres à processus et à archéopyle apical*. BRGM Éditions, Orléans.

Fischer, J.-C. (2003). Invertébrés remarquables du Callovien inférieur de la Voulte-sur-Rhône (Ardèche, France). *Ann. Paléont.*, 89(4), 223–252.

Flörke, O.W., Jones, J.B., Segnit, E.R. (1973). The genesis of hyalite. *N. Jahrb. Miner. Monat.*, 2, 82–89.

Freytet, P. (1984). Les sédiments lacustres carbonatés et leurs transformations par émersion et pédogenèse. Importance de leur identification pour les reconstitutions paléogéographiques. *Bull. Centr. Rech. Explor. Prod. Elf-Aguitaine*, 8, 223–247.

Freytet, P. and Plaziat, J.-P. (1965). Importance de constructions algaires dues à des Cyanophycées dans les formations continentales du Crétacé supérieur et de l'Éocène du Languedoc. *Bull. Soc. Géol. Fr.*, 7, 679–694.

Freytet, P., Lange-Badré, B., Barrier, P., Gand, G., Montenat, C. (2003). La fossilisation des empreintes de pattes et autres traces biologiques. *Le Naturaliste Vendéen*, 3, 61–62.

Fric, A. (1884). O zkamelé babce (*Anomalites fugitivus* Fr.). *Vesmir*, 14, 157–158.

Friis, E.M., Crane P.R., Pedersen, K.R. (2011). *The Early Flowers and Angiosperm Evolution.* Cambridge University Press, Cambridge.

Gall, J.-C. (1990). Les voiles microbiens, leur contribution à la fossilisation des organismes au corps mou. *Lethaia*, 23, 21–28.

Gall, J.-C. (1995). Biofilms et mattes microbiennes : leur contribution à la sédimentogenèse. *C.R. Acad. Sci. Paris*, 321, 823–835.

Gallois, R.W. (2009). The origin of the clay-with-flints: The missing link. *Geosci. South-West Engl.*, 12, 153–161.

Gand, G. (1993). La palichnofaune de vertébrés tétrapodes du bassin permien de Saint-Affrique (Aveyron) : comparaisons et conséquences stratigraphiques. *Géol. France*, 1, 41–56.

Gand, G., Garric, J., Schneider, J., Sciau, J., Walter, H. (1996). Biocoenoses à méduses du Permien français (Bassin de Saint-Affrique, Massif Central). *Geobios*, 29(4), 379–400.

Gand, G., Garric, J., Lapeyrie, J. (1997). Biocénoses à triopsidés (Crustacea, Branchiopoda) du Permien du bassin de Lodève (France). *Geobios*, 30(5), 673–700.

Gand, G., Garric, J., Demathieu, G., Ellenberger, P. (2000). La palichnofaune de vertébrés tétrapodes du Permien Supérieur du Bassin de Lodève (Languedoc-France). *Palaeovertebrata*, 29, 1–82.

Gand, G., Demathieu, G., Montenat, C. (2007). Les traces de pas d'amphibiens, de dinosaures et autres reptiles du Mésozoïque français : inventaire et interpretations. *Palaeovertebrata*, 35, 1–149.

Garcia, G. (2000). Diversité des coquilles minces d'œufs fossiles du Crétacé supérieur du sud de la France. *Geobios*, 33(1), 113–126.

Garcia, G. and Vianey-Liaud, M. (2001). Dinosaur eggshells as biochronological markers in Upper Cretaceous continental deposits. *Palaeogeogr., Palaeoclim., Palaeoecol.*, 169, 153–164.

Gardin, S., Krystyn, L., Richoz, S., Bartolini, A., Galbrun, B. (2012). Where and when the earliest coccolithophores ? *Lethaia,* 45(4), 507–523.

Gaubert, P. (1904). Sur des cristaux de vivianite produite aux dépens d'ossements. *Bull. Soc. fr. Minér.*, 27(7), 212–216.

Gaudant, J., Nel, A., Nury, D., Véran, M., Carnevale, G. (2018). The uppermost Oligocene of Aix-en-Provence (Bouches-du-Rhône, Southern France): A Cenozoic brackish subtropical Konservat-Lagerstätte, with fishes, insects and plants. *C.R. Palevol*, 17, 460–478.

Geigl, E.M. (2008). Palaeogenetics of cattle domestication: Methodological challenges for the study of fossil bones preserved in the domestication centre in Southern Asia. *C.R. Palevol.*, 7, 99–112.

Girard, V., Schmidt, A.R., Saint Martin S., Struwe, S., Perrichot, V., Saint Martin, J.P., Breton, G., Néraudeau, D. (2008). Exceptional preservation of marine diatoms in upper Albian amber. *PNAS*, 105, 17426–17429.

Girard, V., Schmidt, A.R., Struwe, S., Perrichot, V., Breton, G., Néraudeau, D. (2009). Taphonomy and palaeoecology of mid-Cretaceous amber preserved microorganisms from southwestern France. *Geodiversitas*, 31, 152–163.

Giresse, P. and Martzluff, M. (2016). Les calcrètes palustres (tuïres) du Pliocène supérieur de la plaine du Roussillon. Pierres monumentales d'usage historique ancien. *Géologie de la France*, 1, 7–25.

Godefroit, P. and Leduc, T. (2008). La conservation des ossements fossiles : le cas des Iguanodons de Bernissart. *CeROArt*, 2.

Golubovskaya, E.V. (2001). Facies and geochemical features of the Iron Ore Complex of the Kertch Peninsula. *Lithol. Miner. Res.*, 36(3), 224–235.

Gougeon, R., Néraudeau, D., Dabard, M.-P., Pierson-Wickmann, A.C., Polette, F., Poujol, M., Saint Martin, J.-P. (2018). Trace fossils from the Brioverian (Ediacarian-Fortunian) in Brittany (NW France). *Ichnos*, 25(1), 11–24.

Goy, G., Noël, D., Busson, G. (1979). Les conditions de sédimentation des schistes-carton (Toarcien inf.) du bassin de Paris déduites de l'étude des nannofossiles calcaires et des diagraphies. *Doc. Lab. Géol. Fac. Sci. Lyon*, 75, 33–57.

Green, R.E., Briggs, A.W., Krause, J., Prüfer, K., Burbano, H.A., Siebauer, M., Lachmann, M., Pääbo, S. (2009). The Neandertal genome and ancient DNA authenticity. *The EMBO J.*, 28(17), 2494–2502.

Guex, J. (1972). Répartition biostratigraphique des ammonites du Toarcien moyen de la bordure sud des Causses (France) et révision des ammonites décrites et figurées par Monestier (1931). *Eclogae Geol. Helv.*, 65(3), 611–645.

Iskandar, D. (1990). La diatomite miocène de la montagne d'Andance, carrière de Saint-Bauzile (Ardèche, France). Etude palynologique, écostratigraphique, paléoclimatologie. *Doc. Lab. Géol. Fac. Sci. Lyon*, 110, 1–173.

Jacob, J., Paris, F., Monod, O., Miller, M.A., Tang, P., George, S.C., Bény, J.-M. (2007). New insights into the chemical composition of chitinozoans. *Organ. Geochem.*, 38(10), 1782–1788.

Jattiot, R., Trincal, V., Moreau, J.-D., Brocard, A. (2015a). *Guide des ammonites pyriteuses. Toarcien moyen et supérieur des Causses (Lozère – France)*. Les Editions du Piat, Paris.

Jattiot, R., Brayard, A., Fara, E., Charbonnier, S. (2015b). Gladius-bearing coleoids from the Upper Cretaceous Lebanese Lagerstätten: Diversity, morphology, and phylogenetic implications. *J. Paleont.*, 89(1), 148–167.

Jattiot, R., Coquel-Poussy, N., Kruta, I., Rouget, I., Rowe, A.J., Moreau, J.D. (2024). The first gladius-bearing coleoid cephalopods from the lower Toarcian "Schistes Cartons" Formation of the Causses Basin (southeastern France). *PeerJ.*, 12, e16894.

Jenny, J. (1988). Carte géologique du Maroc au 1/100 000. Feuille Azizal (Haut Atlas Central). Mémoire explicatif. *Notes Mém. Serv. Géol. Maroc*, 339, 1–104.

Jouault, C. and Nel, A. (2022). A new fossil representative of the scorpionfly subfamily Panorpinae (Mecoptera, Panorpidae) from the Miocene of France. *Palaeoentomology*, 5(2), 142–145.

Jouault, C., Legendre, F., Perrichot, V., Condamine, F., Nel, A. (2022a). Redescription of *Litholingia rhora* Ren, 2002 (Neuroptera: Gramma-lingiidae) from the Middle Jurassic of Daohugou. *Palaeoentomology*, 5(1), 020–026.

Jouault, C., Nam, G.S., Nel, A. (2022b). The first Grammalingiidae (Polyneoptera: Reculida) from the Upper Triassic Amisan Formation of South Korea. *Ann. Paléont.*, 108, 102558.

Jouault, C., Pouillon, J.C., Nel, A. (2022c). Beautiful wing coloration pattern in a new doratomantispine species (Neuroptera: Mantispidae) from the mid-Cretaceous Burmese amber. *Zootaxa*, 5100(3), 390–400.

Jouault, C., Tischlinger, H., Henrotay, M., Nel, A. (2022d). Wing coloration patterns in the Early Jurassic dragonflies as potential indicator of increasing predation pressure from insectivorous reptiles. *Palaeoentomology*, 5(4), 305–318.

Kerp, H. (2018). Organs and tissues of Rhynie chert plants. *Phil. Trans. Roy. Soc. B, Biol. Sci.*, 373(1739), 20160495.

Kohn, M.J., Schoeninger, M.J., Barker, W.W. (1999). Altered states: Effects of diagenesis on fossil tooth chemistry. *Geochim. Cosmochim. Acta*, 63(18), 2737–2747.

Kokinos, J.P., Eglinton, T.I., Goñi, M.A., Boon, J.J., Martoglio, P.A., Anderson, D.M. (1998). Characterization of a highly resistant biomacromolecular material in the cell wall of a marine dinoflagellate resting cyst. *Org. Geochem*, 28(5), 265–288.

Kosakevitch, A. (1983). Étude morphogénétique des textures de remplacement et de remplissage dans les chapeaux de fer. Report, Bureau de Recherches Géologiques et Minières, Orléans.

Krings, M., Stone, A., Schmitz, R.W., Krainitzki, H., Stoneking, M., Pääbo, S. (1997). Neandertal DNA sequences and the origin of modern humans. *Cell*, 90(1), 19–30.

Krings, M., Geisert, H., Schmitz, R.W., Krainitzki, H., Pääbo, S. (1999). DNA sequence of the mitochondrial hypervariable region II from the Neandertal type specimen. *Proceed. Nat. Acad. Sci.*, 96(10), 5581–5585.

Krumbein, W.E. (1978). Algal mats and their lithification. In *Environmental Biogeochemistry and Geomicrobiology: The Aquatic Environment*, Krumbein, W.E. (ed.). Ann Arbor Science, Ann Arbor.

Kuhlwilm, M., Gronau, I., Hubisz, M.J. et al. (2016). Ancient gene flow from early modern humans into Eastern Neanderthals. *Nature*, 530, 429–433.

Lacroix, A. (1910). *Minéralogie de la France et de ses colonies*, Volume 3, Part 1. Librairie Polytechnique, Paris.

Lak, M., Néraudeau, D., Nel, A., Cloetens, P., Perrichot, V., Tafforeau, P. (2008). Phase contrast X-ray synchrotron imaging: Opening access to fossil inclusions in opaque amber. *Microscopy & Microanalysis*, 14, 251–259.

Laroche, J., Guervin, C., Le Coq, C. (1989). Phénomènes de pétrification réalisés in vitro, II – Premiers stades de la silicification. *Lettres Botaniques*, 136, 267–279.

Laveine, J.P. (2007). La formation du charbon. Au début, il y avait la forêt. Exhibition Catalogue, Muséum d'Histoire Naturelle de Lille, Lille.

Le Couls, M. (2017). Les paléoflores du Jurassique moyen et supérieur (Bathonien-Oxfordien) du bassin parisien et de ses marges. PhD Thesis, Université de Rennes 1, Rennes.

Le Couls, M., Courville, P., Moreau, J.-D. (2015). La paléoflore bathonienne (Jurassique moyen) des silex à plantes d'Arc-en-Barrois (Haute-Marne, Nord-Est de la France). *Journées scientifiques du Groupe Français du Jurassique*, Dijon.

Lebrun, P. (2013). Les fossiles des diatomites du Miocène supérieur de Saint-Bauzile, Ardèche. *Fossiles*, 15, 16–22.

Lebrun, P. (2015). En bref : actualités paléontologiques. *Fossiles*, 23, 65.

Legendre, S., Sigé, B., Astruc, J.G. et al. (1997). Les phosphorites du Quercy : 30 ans de recherche. Bilan et perspectives. *Geobios*, 30, 331–345.

Lemierre, A., Folie, A., Bailon, S., Robin, N., Laurin, M. (2021). From toad to frog, a CT-based reconsideration of *Bufo servatus*, an Eocene anuran mummy from Quercy (France). *J. Vert. Paleont.*, 41(3), e1989694.

Lemierre, A., Gendry, D., Poirier, M.M., Gillet, V., Vullo, R. (2023). The oldest articulated ranid from Europe: A *Pelophylax* specimen from the lowest Oligocene of Chartre-de-Bretagne (N.W. France). *J. Vert. Paleont.*, 42(4), e2191663.

Li, Q., Gao, K.Q., Vinther, J., Shawkey, M.D., Clarke, J.A., D'Alba, L., Meng, Q., Briggs, D.E.G., Prum, R.O. (2010). Plumage color patterns of an extinct dinosaur. *Science,* 327, 1369–1372.

Li, Q., Gao, K.Q., Meng, Q., Clarke, J.A., Shawkey, M.D., D'Alba, L., Pei, R., Ellison, M., Norell, M.A. Vinther, J. (2012). Reconstruction of *Microraptor* and the evolution of iridescent plumage. *Science*, 335(6073), 1215–1219.

Limondin-Lozouet, N., Dabkowski, J., Antoine, P., Bahain, J.J., Carbonel, P., Ghaleb, B. (2012). Les formations de tuf de Condat-sur-Vézère et de la vallée du Coly. *Quaternaire Continental d'Aquitaine, AFEQ*, 88–100.

Lin, C.Y. (2020). The microbially driven formation of siderite in salt marsh sediments. *Geobiology*, 18, 207–224.

Lomax, D.R. and Racay, C.A. (2012). A long mortichnial trackway of *Mesolimulus walchi* from the Upper Jurassic Solnhofen Lithographic Limestone near Wintershof, Germany. *Ichnos*, 19(3), 175–183.

Lopez, M., Gand, G., Garric, J., Körner, F., Schneider, J. (2008). The playa environments of the Lodève Permian basin (Languedoc-France). *J. Iberian Geol.*, 34, 29–56.

Lynne, B.Y. and Campbell, K.A. (2004). Morphologic and mineralogic transitions from Opal-A to Opal-CT in low-temperature siliceous sinter diagenesis, Taupo Volcanic Zone, New Zealand. *J. Sedim. Res.*, 74(4), 561–579.

Martín-Closas, C. and Gomez, B. (2004). Taphonomie des plantes et interprétations paléoécologiques. Une synthèse. *Geobios*, 37, 65–88.

Masson-Delmotte, V. and Chapellaz, J. (2002). Au cœur de la glace, les secrets du climat. *Bull. Union Physic.*, 96, 1695–1707.

Maurel, C. (1964). Types de réactions d'oxydation observés au cours de l'analyse thermique différentielle, dans l'air, de minéraux sulfurés et arséniés de Fer, Co, Ni, Cu, Zn, Ag et Pb. *Bull. Soc. fr. Minér. Cristallog.*, 87(3), 377–385.

Mazin, J.-M., Pélissié, T., Hantzpergue, P., Lézin, C. (2013). Gisement fossile et ichnologique sur la marge Nord-Est du Bassin d'Aquitaine (Quercy). Livret-guide de l'excursion. *Strata*, 15, 1–42.

Ménillet, F. (1987). Les meulières du Bassin de Paris (France) et les faciès associés. Rôle des altérations supergènes néogènes à quaternaire ancien dans leur génèse. PhD Thesis, Université Louis Pasteur, Strasbourg.

Merle, D. (2022). Motifs et couleurs fossiles des coquillages. *Espèces*, 45, 58–65.

Monestier, J. (1931). Ammonites rares ou peu connues et ammonites nouvelles du Toarcien moyen de la région sud-est de l'Aveyron. *Mém. Soc. Géol. Fr.*, 15, 1–79.

Monestier, J. (1934). Ammonites du Domérien de la région du sud-est de l'Aveyron et de quelques régions de la Lozère à l'exclusion des Amalthéidés. *Mém. Soc. Géol. Fr.*, 23, 1–102.

Moreau, J.-D., Cloetens, P., Gomez, B., Daviero-Gomez, V., Néraudeau, D., Lafford, T., Tafforeau, P. (2014). Multiscale 3D virtual dissections of 100-million-year-old flowers using X-ray synchrotron micro- and nanotomography. *Microscopy & Microanalysis*, 20, 305–312.

Moreau, J.-D., Néraudeau, D., Tafforeau, P., Dépré, É. (2015). Study of the histology of leafy axes and male cones of *Glenrosa carentonensis* sp. nov. (Cenomanian Flints of Charente-Maritime, France) using synchrotron microtomography linked with palaeoecology. *PLoS ONE,* 10(8), e0134515.

Moreau, J.-D., Néraudeau, D., Platel, J.P., Ravon, A.L. (2016a). Les silex fossilifères (invertébrés marins et plantes terrestres) du Crétacé supérieur de Claix (Charente). *Ann. Paléont.*, 102, 103–116.

Moreau, J.-D., Gomez, B., Coiffard, C., Daviero-Gomez, V., Tafforeau, P., Néraudeau, D. (2016b). *Verneda hermaphroditica* gen. & sp. nov.: A new flower head from the early Late Cretaceous (middle Cenomanian) of southeastern France. *Taxon*, 65(4), 823–838.

Moreau, J.-D., Néraudeau, D., Tafforeau, P., Dépré, É. (2017a). Synthèse sur la diversité des préservations végétales du site d'Archin-geay-Les Nouillers (France) : un Konservat-Lagerstätte pour les flores de l'Albien-Cénomanien. *Ann. Paléont.*, 103, 165–171.

Moreau, J.-D., Néraudeau, D., Perrichot, V., Tafforeau, P. (2017b). 100-million-year-old conifer tissues from the mid-Cretaceous amber of Charentes (western France) revealed by synchrotron microtomography. *Ann. Bot.,* mcw225.

Moreau, J.-D., Néraudeau, D., Vullo, R., Abit, D., Mennecart, B., Schnyder, J. (2017c). Late Jurassic dinosaur footprints from Chassiron-La Morelière (Oléron Island, western France). *Palaeobiol. Palaeoenvir.*, 97, 773–789.

Moreau, J.-D., Néraudeau, D., Benoit, R.-A., Dépré, É, Le Couls, M., Fernandez, V. (2019a). Exceptional preservation of Cretaceous biota (plants and invertebrates) in flint-bearing alterites from Moragne (western France). *Palaeont. Electr.*, 22.3.66, 1–14.

Moreau, J.-D., Trincal, V., Nel, A., Simon-Coinçon, R., Sallé, V., Le Couls, M., Néraudeau, D., Fernandez, V. (2019b). Hidden termite coprolites revealed by synchrotron microtomography inside Eocene-Oligocene filled wood-borings from the Malzieu Basin, Lozère, southern France. *Lethaia*, 53, 106–117.

Moreau, J.-D., Philippe, M., Néraudeau, D., Dépré, É., Le Couls, M., Fernandez, V., Beurel, S. (2021). Paleohistology of the Cretaceous resin-producing conifer *Geinitzia reichenbachii* using X-ray synchrotron microtomography. *Amer. J. Botany*, 108(9), 1745–1760.

Moreau, J.-D., Vullo, R., Charbonnier, S., Jattiot, R., Trincal, V., Néraudeau, D., Fara, E., Baret, L., Garassino, A., Gand, G., Lafaurie, G. (2022). Konservat-Lagerstätten from the Upper Jurassic lithographic limestone of the Causse Méjean (Lozère, southern France): Palaeontological and palaeoenvironmental synthesis. *Geol. Mag.*, 159(5), 761–781.

Moreau, J.-D., Michelin, A., Fara, E., Gand, G., Galtier, J., Puech, G., Fouché, S. (2024). Ichnofossils and body fossils from the Permian of the Sorgue Valley (Saint-Affrique Basin, southern France): Palaeoenvironmental implications. *Histor. Biol.,* 36, 82–95.

Morozewicz, M.J. (1924). La grodnolite, phosphate colloïdal de calcium. *Bull. Minéral.,* 47(3–4), 46–48.

Moussu, R. (1959). La minéralisation cuprifère d'âge atlasique du Haut Atlas central et oriental (Maroc). In *20ème Congr. Géol. Internat.*, 279–288.13.

Munnecke, A. and Servais, T. (1996). Scanning electron microscopy of polished, slightly etched rock surfaces: A method to observe palynomorphs *in situ. Palynology*, 20, 163–176.

Mustoe, G.E. (2018). Mineralogy of non-silicified fossil wood. *Geosciences*, 8(3), 1–17.

Mustoe, G.E. and Smith, E.T. (2023). Timing of opalization at Lightning Ridge, Australia: New evidence from opalized fossils. *Minerals*, 13, 1471.

Nel, A. and Jouault, C. (2022). New grasshoppers (Orthoptera: Elcanidae, Locustopsidae) from the Lower Cretaceous Crato Formation suggest a biome homogeneity in Central Gondwana. *Hist. Biol.*, 34(10), 2070–2078.

Nel, A., Gill, G.A., Nury, D. (1987). Découverte d'empreintes attribuables à des Coelenthérés Siphonophores chondrophorides dans l'Oligocène de Provence. *C.R. Acad. Sci. Paris*, 2(305), 637–641.

Nel, A., De Ploëg, G., Dejax, J. et al. (1999). Un gisement sparnacien exceptionnel à plantes, arthropodes et vertébrés (Éocène basal, MP7) : Le Quesnoy (Oise, France). *C.R. Acad. Sci.*, 329(1), 65–72.

Néraudeau, D. (2015). Origine géologique des silex à plantes de Torsac (Charente). *Ann. Soc. Sci. Nat. Charente-Maritime*, 10(6), 583–591.

Néraudeau, D., Perrichot, V., Dejax, J., Masure, E., Nel, A., Philippe, M., Moreau, P., Guillocheau, F., Guyot, T. (2002). Un nouveau gisement à ambre insectifère et à végétaux (Albien terminal probable) : Archingeay (Charente-Maritime, France). *Geobios*, 35, 233e240.

Néraudeau, D., Allain, R., Ballèvre, M. et al. (2012). The Hauterivian-Barremian lignitic bone bed of Angeac (Charente, south-west France): Stratigraphical, palaeobiological and palaeogeographical implications. *Cret. Res.*, 37, 1–14.

Néraudeau, D., Mazan, P., Vullo, R. (2013). *Fossiles de la Préhistoire charentaise*. Le Croît Vif, Paris.

Néraudeau, D., Dabard M.-P., El Albani A., Gougeon R., Mazurier A., Pierson-Wickmann A.C., Poujol M., Saint Martin, J.P., Saint Martin, S. (2018). First evidence of Ediacaran-Fortunian elliptical body fossils in the Brioveran series of Brittany (NW France). *Lethaia*, 51(4), 513–522.

Néraudeau, D., Vullo, R., Bénéfice, P. et al. (2019). The paralic Albian-Cenomanian Puy-Puy Lagerstätte (Aquitaine Basin, France): An overview and new data. *Cret. Res.*, 111, 104124.

Néraudeau, D., Philippe, M., El Albani, A., Moreau, J.-D., Polette, F., Saint Martin, J.P. (2022). Taphonomie et minéralisation de bois carbonatisés du Purbeckien de l'Usine, à Cherves-Richemont (Charente). *Geodiversitas*, 44(24), 665–682.

Néraudeau, D., Coutret, B., Gendry, D., Lesage, B., Loi, A., Poujol, M. (2023). Elliptical body fossils from the Fortunian (Early Cambrian) of Normandy (NW France). *C.R. Palevol*, 22(19), 439–453.

Noël, D., Busson, G., Cornée, A., Bodeur, Y., Mangin, A.-M. (1994a). Contribution fondamentale des Coccolithophoridées à la constitution des calcaires fins du Jurassique moyen et supérieur. *Geobios*, 27(3), 701–721.

Noël, D., Busson, G., Cornée, A., Bodeur, Y., Mangin, A.-M. (1994b). Le nannoplancton calcaire et la formation des alternances calcaires-marnes dans le Lias des bassins de Marches-Ombrie (Italie). *Riv. It. Paleont. Strat.*, 99(4), 515–550.

Norman, D.B. (1980). On the ornithischian dinosaur *Iguanodon bernissartensis* from the Lower Cretaceous of Bernissart (Belgium). *Mém. Inst. Roy. Sci. Nat. Belgique*, 178, 1–105.

Orlando L., Ginolhac, A., Zhang, G. et al. (2013). Recalibrating *Equus* evolution using the genome sequence of an early Middle Pleistocene horse. *Nature*, 499, 74–78.

Paris, F. (2006). Chitinozoans: A fascinating and mysterious microfossil-group. Technical-card and direction for use. doi: 10. 13140/RG.2.1.2462.7682.

Paris, F. and Nõlvak, J. (1999). Biological interpretation and paleobiodiversity of a cryptic fossil group: The “chitinozoan animal”. *Geobios*, 32(2), 315–324.

Paris, F., Girard, C., Feist, R., Winchester-Seeto, T. (1996). Chitinozoan bioevent in the Frasnian-Famennian boundary beds at La Serre (Montagne Noire, Southern France). *Palaeogeogr., Palaeoclimatol., Palaeoecol.*, 121, 131–145.

Paris, F., Achab, A., Asselin, E. (2004). Chitinozoans. In *The Great Ordovician Biodiversification Event*, Webby, B., Paris, F., Droser., Percival, I. (eds). Columbia University Press, New York.

Pélissié, T., Sige, R., Rey, R. (2006). 30 millions d’années de biodiversité dynamique dans le paléokarst du Quercy. *Strata*, 1(13), 1–284.

Penney, D. (2010). *Biodiversity of Fossils in Amber from the Major World Deposits*. Siri Scientific Press, Manchester.

Penney, D. (2013). Absence of ancient DNA in sub-fossil insect inclusions preserved in “Anthropocene” Colombian copal. *PLOS One*, 8(9), e73150.

Penney, D. and Preziosi, R.F. (2010). On inclusions in subfossil resins (copal). In *Biodiversity of Fossils in Amber from the Major World Deposits*, Penney, D. (ed.). Siri Scientific Press, Manchester.

Perrichot, V. (2005). Environnements paraliques à ambre et à végétaux au Crétacé nord-aquitain (Charentes, Sud-Ouest de la France). *Mém. Géosci. Rennes*, 118, 1–310.

Perrichot, V., Néraudeau, D., Azar, D., Menier, J.J., Nel, A. (2002). A new genus and species of fossil mole cricket in the Lower Cretaceous amber of Charente-Maritime, SW France (Insecta: Orthoptera: Gryllotalpidae). *Cret. Res.*, 23, 307–314.

Perrichot, V., Marion, L., Néraudeau, D., Vullo, R., Tafforeau, P. (2008). The early evolution of feathers: Fossil evidence from Cretaceous amber of France. *Proceed. Roy. Soc. London, Biol. Sci.*, 275, 1197–1202.

Perrier, V. and Charbonnier, S. (2014). The Montceau-les-Mines Lagerstätte (Late Carboniferous, France). *C.R. Palevol*, 13(5), 353–367.

Pestrea, S. and Saint Martin, J.-P. (2002). La microflore de diatomées du Messinien de Gibellina (Sicile occidentale, Italie). *Geodiversitas*, 24(3), 585–610.

Philippot, A. (1950). Les Graptolites du Massif armoricain. Étude stratigraphique et paléontologique. *Mém. Soc. Géol. Minéral. Bretagne*, 11, 1–295

Piçarra, J.M., Robardet, M., Bourahrouh A., Paris, F., Pereira, Z., Le Menne, J., Gourvennec, R., Oliveira, T., Lardeux, H. (2002). Le passage Ordovicien-Silurien et la partie inférieure du Silurien (Sud-Est du Massif armoricain, France). *C.R. Géoscience*, 334(16), 1177–1183.

Pignatelli, I., Giuliani, G., Morlot, C., Salsi, L., Martayan, G. (2022). Les curiosités minéralogiques de l'émeraude colombienne. In *Émeraude, tout un monde !,* Giuliani, G. (ed.). Editions du Piat, Paris.

Poinar, H.N. and Stankiewicz, B.A. (1999). Protein preservation and DNA retrieval from ancient tissues. *Proc. Natl. Acad. Sci. USA*, 96, 8426–8431.

Polette, F. (2019). Les assemblages palynologiques continentaux du Crétacé Inférieur de France (Tithonien-Cénomanien) : paléoenvironnements, paléoclimats, stratigraphie, et taxonomie. PhD Thesis, Université Rennes 1, Rennes.

Polette, F., Batten, Néraudeau, D. (2018). Re-examination of the palynological content of the Lower Cretaceous deposits of Angeac, Charente, south-west France: Age, palaeoenvironment and taxonomic determinations. *Cretaceous Research*, 90, 204-221.

Pons, T. (2015). Caractérisation des oxy-hydroxydes de fer et des éléments associés (S, Se, As, Mo, V, Zr) dans les environnements redox favorables aux gisements d'uranium. PhD Thesis, Université Paris 11, Paris.

Porpino Lemos, V., Lima da Costa, M., Lima Lemos, R., Gomez de Faria, M.S. (2007). Vivianite and siderite in lateritic iron: An example of bioreduction. Quim. *Nova*, 30(1), 36–40.

Pouclet, A. (2019). Deuxième sortie géologique de l'AVG sur le Jurassique du littoral vendéen, de St-Nicolas-sur-Jard vers Jard-sur-Mer. Guidebook, Association Vendéenne de Géologie, La Roche-sur-Yon.

Pouech, J., Amiot, R., Lécuyer, C., Mazin J.-M., Martineau, F., Fourel, F. (2014). Oxygen isotope composition of vertebrate phosphates from Cherves-de-Cognac (Berriasian, France): Environmental and ecological significance. *Palaeogeogr., Palaeoclim., Palaeoecol.*, 410, 290–299.

Powell, J.H. (2012). Early diagenesis of Late Cretaceous chalk-chert-phosphorite hardgrounds in Jordan: Implications for sedimentation on a Coniacian-Campanian pelagic ramp. *GeoArabia*, 17(4), 17–38.

Prüfer, K., Racimo, F., Patterson, N. et al. (2014). The complete genome sequence of a Neanderthal from the Altai Mountains. *Nature*, 505(7481), 43–49.

Püttmann, T. and Mutterlose, J. (2021). Paleoecology of Late Cretaceous aoccolithophores: Insights from the Shallow-Marine Record. *Paleoceanogr. and Paleoclimatol.*, 36(3).

Ragazzi, E., Roghi, G., Giaretta, A., Gianolla, P. (2003). Classification of amber based on thermal analysis. *Thermochim. Acta*, 404, 43–54.

Raynaud, D. and Lorius, C. (2004). Climat et atmosphère : la mémoire des glaces. *C.R. Géoscience*, 336(7–8), 647–656.

Reich, D., Green, R.E., Kircher, M. et al. (2010). Genetic history of an archaic hominin group from Denisova Cave in Siberia. *Nature*, 468, 1053–1060.

Renault, B. and Zeiller, R. (1890). *Etudes sur le terrain houiller de Commentry. Livre deuxième : flore fossile*. Imprimerie Théolier, Saint-Etienne.

Robertson, R.S. and Scott, D.C. (1990). Geology of the Coober Pedy precious stones field. Results of investigations, 1981-86. *Geol. Surv. South Australia, Report of Investigations*, 56, 1–55.

Robin, N., Velasquez, M., Boura, A., Garcia, G., Jauvion, C., Boiteau, J.M., Gomez, B., Daviero-Gomez, V., Valentin, X. (2018). The oldest shipworms (Bivalvia, Pholadoidea, Teredinidae) preserved with soft parts (western France): Insights into the fossil record and evolution of Pholadoidea. *Palaeontology*, 61(6), 905–918.

Robin, N., Noirit, F., Chevrinais, M., Clément, G., Olive, S. (2022). Vertebrate predation in the Late Devonian evidenced by bite traces and regurgitations: Implications within an early tetrapod freshwater ecosystem. *Papers in Palaeontology*, 8(4), e1460.

Roghi, G., Ragazzi, E., Gianolla, P. (2006). Triassic amber of the Southern Alps (Italy). *Palaios*, 21, 143–154.

Roth, P.H. (1984). Preservation of calcareous nannofossils and fine-grained carbonate particles in Mid-Cretaceous sediments from the Southern Angola basin. Initial Reports of the Deep Sea Drilling Project, 75, U.S. Government Printing Office, Washington, D.C., 651–655.

Saint Martin, J.P., Métais, G., Saint Martin, S., Sen, S. (2017). La diatomite du Coiron et son lagerstätte. *Géochronique*, 141, 57–69.

Santos, A.A., Xiao, L., Labandeira, C.C., Néraudeau, D., Dépré, É., Moreau, J.-D., Perrichot, V., Wappler, T. (2022). Plant-insect interactions from the mid-Cretaceous at Puy-Puy (Aquitaine Basin, western France) indicates preferential herbivory for angiosperms amid a forest of ferns, gymnosperms, and angiosperms. *Bot. Lett.*, 169(4), 568–587.

Sargent Bray, P., Anderson, K.B. (2009). Identification of Carboniferous (320 million years old) classic amber. *Science*, 326(5949), 132–134.

Schlüter, T. and von Gnielinski, F. (1987). The East African Copal, its geologic, stratigraphic, palaeontologic significance and comparison with fossil resins of similar age. *Occ. Pap. Nat. Mus. Tanzania*, 8, 1–32.

Schoeninger, M.J., Hallin, K., Reeser, H., Valley, J.W., Fournelle, J. (2003). Isotopic alteration of mammalian tooth enamel. *Int. J. Osteoarchaeol.*, 13, 11–19.

Schopf, J.M. (1975). Modes of fossil preservation. *Rev. Palaeobot. Palynol.*, 20, 27–53.

Schweigert, G., Maxwell, E., Dietl, G. (2016). First record of a true mortichnium produced by a fish. *Ichnos*, 23(1–2), 71–76.

Scott, A.C. (2000). The Pre-Quaternary history of fire. *Palaeogeogr., Palaeoclim., Palaeoecol.*, 164, 281–329.

Seilacher, A. (1970). Begriff und Bedeutung der Fossil-Lagerstätten. *N. Jb. Geol. Palaont. Mh.*, 1, 34–39.

Seilacher, A. (2008). Biomats, biofilms, and bioglue as preservational agents for arthropod trackways. *Palaeogeogr., Palaeoclim., Palaeoecol.*, 270(3–4), 252–257.

Shannon, E. (1921). Description of vivianite encrusting a fossil tusk from gold placers of Clearwater County, Idaho. *Proceed. U.S. N. Mus.*, 59(2375), 415–417.

Smallwood, A., Thomas, P.S., Ray, A. (2008). Comparative analysis of sedimentary and volcanic precious opals from Australia. *J. Austr. Ceram. Soc.*, 44(2), 17–22.

Smith, K.T., Schaal, S.F.K., Habersetzer, J. (2018). *Messel – An Ancient Greenhouse Ecosystem*. Schweizerbart, Stuttgart.

Soriano, C., Archer, M., Azar, D. et al. (2010). Synchrotron X-ray imaging of inclusions in amber. *C.R. Palevol*, 9(6–7), 361–368.

Stock, C. (1992). Rancho La Brea: A record of Pleistocene life in California (Revised by J.M. Harris). *Nat. Hist. Mus. Los Angeles Count.*, 37, 1–113.

Szewczyk, L., Vennin, E., Moreau, J.-D., Gand, G., Vérolet, M., Klee, N. Fara, E. (2020). Tracking dinosaurs in coarse-grained sediments from the Upper Triassic of Ardèche (southern France). *Palaeogeogr., Palaeoclimat., Palaeoecol.*, 35, 447–460.

Tafforeau, P., Boistel, R., Boller, E. et al. (2006). Applications of X-ray synchrotron microtomography for non-destructive 3D studies of paleontological specimens. *Appl. Phys. A Mater. Sci. Process.*, 83, 195–202.

Taylor, E.L., Taylor, T.N., Krings, M. (2009). *Paleobotany: The Biology and Evolution of Fossil Plants*. Academic Press, Burlington.

Thiry, M., Milnes, A.R., Rayot, V., Simon-Coinçon, R. (2006). Interpretation of palaeoweathering features and successive silicifications in the Tertiary regolith of inland Australia. *J. Geol. Soc.*, 1563, 723–736.

Thomas, P. (2008). Un gisement d'hydrocarbures vu de l'intérieur et un trésor du patrimoine géologique français : la mine de bitume de Dallet (Puy-de-Dôme), dite "Mine des Rois" [Online]. Available at: https://planet-terre.ens-lyon.fr/ressource/mine-rois-bitume-Dallet.xml.

Thoral, M. (1946). Cycles géologiques et formations nodulifères de la Montagne Noire. *Nouv. Archiv. Mus. Hist. Nat. Lyon*, 1, 1–63.

Tillyard, M.A. (1922). An insect wing in a crystal of selenite (order Orthoptera). *Rec. Geol. Surv. New South Wales*, 205–207.

Van der Valk, T., Pečnerová, P., Díez-del-Molino, D. et al. (2021). Million-year-old DNA sheds light on the genomic history of mammoths. *Nature*, 591, 265–269.

Van Eetvelde, Y. and Dupuis, C. (2005). Etude biostratigraphique et paléoenvironnementale des diatomées pyritisées de l'intervalle Paléocène supérieur-Eocène inférieur du sondage de Saint-Josse (France). *Geobios*, 38, 269–282.

Vénec-Peyré, M.-T. (1981). Les foraminifères et la pollution : étude de la microfaune de la cale du Dourduff (embouchure de la rivière de Morlaix). *Cahiers de Biologie Marine*, 22, 25–33.

Vogt, T. (1984). Les croûtes calcaires : types et genèse – exemples de l'Afrique du Nord et de la France méditerranéenne. PhD Thesis, Université Louis Pasteur, Strasbourg.

Vuillemin, A., Wirth, R., Kemnitz, H. et al. (2019). Formation of diagenetic siderite in modern ferruginous sediments. *Geology*, 47(6), 540–544.

Vuillet, P., Giuliani G., Fischer, J.-C., Chiappero, P.-J. (2002). Les émeraudes de Gachala, Colombie. Historique, genèse et découvertes paléontologiques. *Le Règne Minéral*, 46, 5–18.

Vullo, R. and Néraudeau, D. (2008). Cenomanian vertebrate assemblages from south-western France: A new insight into the European mid-Cretaceous continental fauna. *Cret. Res.*, 29(5–6), 930–935.

Vullo, R., Néraudeau, D., Videt, B. (2003). Un faciès de type falun dans le Cénomanien basal de Charente-Maritime (France). *Ann. Paléont.*, 89(3), 171–189.

Vullo, R., Girard, V., Azar, D., Néraudeau, D. (2010). Mammalian hair in Early Cretaceous amber. *Naturwissenschaften*, 97, 683–687.

Vullo, R., Néraudeau, D., Dépré, É. (2013). Vertebrate remains from the Cenomanian (Late Cretaceous) plant-bearing Lagerstätte of Puy-Puy (Charente-Maritime). *Cret. Res.*, 45, 314–320.

Weller, S. (1907). A report on the Cretaceous paleontology of New Jersey. *Geol. Surv. New Jersey*, 4, 1–871.

Wilbert, J. (1962). Croûtes et encroutements calcaires au Maroc. *Al Awamia*, 3, 175–192.

Willerslev, E. and Cooper, A. (2005). Ancient DNA. *Proc. R. Soc. B.*, 272, 3–16.

Workman, C., Dalen, L., Vartayan, S., Shapiro, B., Kosintsev, P., Sher, A., Gotherstrom, A., Barnes, I. (2011). Population-level genotyping of coat colour polymorphism in woolly mammoth (*Mammuthus primigenius*). *Quatern. Sci. Rev.*, 30, 2304–2308.

Zhang, F., Kearns, S., Orr, P., Benton, M., Zhou, Z., Johnson, D., Xu, X., Wang, X. (2010). Fossilized melanosomes and the colour of Cretaceous dinosaurs and birds. *Nature*, 463(7284), 1075–1078.

List of Figures

Chapter 4

List of Plates

Chapter 1

Chapter 2

Chapter 3

Chapter 4

List of Authors

Bruno DAVID
Anciennement Muséum National d'Histoire Naturelle
Paris
France

Jean-David MOREAU
GEOPS, CNRS
Université Paris-Saclay
Orsay
France

Didier NÉRAUDEAU
Géosciences, CNRS
Rennes
France

Index

C

D

G

H

I, J

L

N

O

T

Printed and bound by CPI Group (UK) Ltd, Croydon, CR0 4YY

13/09/2026

14961073-0001